Current Research in Sports Biomechanics

Medicine and Sport Science

Founder and Editor from 1969 to 1984
E. Jokl, Lexington, Ky.

Vol. 25

Basel · München · Paris · London · New York · New Delhi · Singapore · Tokyo · Sydney

Current Research in Sports Biomechanics

Volume Editors
B. Van Gheluwe, Brussels
J. Atha, Loughborough

114 figures and 6 tables, 1987

Basel · München · Paris · London · New York · New Delhi · Singapore · Tokyo · Sydney

Medicine and Sport Science

Published on behalf of the Research Committee of the International Council of Sport Sciences and Physical Education

Library of Congress Cataloging-in-Publication Data

Current research in sports biomechanics.
(Medicine and sport science; vol. 25)
Includes bibliographies and index.
1. Sports-Physiological aspects. 2. Human mechanics. I. Gheluwe, B. van. II. Atha, John. III. Series. [DNLM: 1. Biomechanics. 2. Sports.
W1 ME649Q v. 25/WE 103 C976]
RC1235.C87 1987 612'.044 87-3049
ISBN 3-8055-4546-0

Drug Dosage

The authors and the publisher have exerted every effort to ensure that drug selection and dosage set forth in this text are in accord with current recommendations and practice at the time of publication. However, in view of ongoing research, changes in government regulations, and the constant flow of information relating to drug therapy and drug reactions, the reader is urged to check the package insert for each drug for any change in indications and dosage and for added warnings and precautions. This is particularly important when the recommended agent is a new and/or infrequently employed drug.

Printed in Switzerland by Thür AG Offsetdruck, Pratteln
ISBN 3-8055-4546-0

Contents

Preface

The emergence of biomechanics as a modern discipline may reasonably be said to date from the time of the First International Biomechanics Seminar at Zürich in 1967. By that date, a general awareness had developed that biomechanics was no longer the esoteric domain of a few scattered pioneers, but an applied science with an honourable history and a respected capacity for attracting the sustained attention of research scientists throughout the developed world. Based firmly upon the biological and applied mathematical sciences, biomechanics counts among its devotees anatomists and physiologists, mathematicians and physicists, sports scientists and ergonomists, space scientists and medical practitioners. Its has its focus on the properties of materials and the characteristics of biological tissues; on the testing of consumer products, protective clothing and equipment; on the movements of animals and the flight of birds; on the activities of the disabled, the behaviour of industrial workers and the performance of elite athletes; on safety, workplace design, therapeutics and medical diagnosis.

The persistent quest for the ultimate performance, the holy grail for athlete and coach, has long fuelled the drive to apply sound quantitative techniques to the analysis of sports performance, and this has made sports biomechanics a lusty limb of its parent discipline. At the same time, the needs of the ordinary individual undertaking recreational activities have not been entirely neglected, and benchmarks of broad scientific interest as well as of narrow specialist application have been established. Furthermore, the recent and burgeoning commercial interest in sports technology (the leisure and sports industry in Britain has a turnover of over one billion pounds sterling) has led to the allocation of some resources for scientific research in commercially profitable areas. Such forces as these are acting to carry sports biomechanics to a healthy maturity.

Nonetheless much development has still to occur. In particular, few textbooks are yet in print, and these are characteristically written for readers with modest levels of numeracy. Journal or conference papers, although often excellent, are typically short and tightly focussed. This publication seeks to fill the gap between them by providing an opportunity for a select group of rising young individuals and established figures to engage in a fuller discussion of topics currently under research scrutiny. In the following pages, attention is given to the biomechanics of particular events such as running or gymnastics, to analytical techniques such as the measurement of power or the quantification of dynamic muscle activity, and to the pragmatic process of combining quantitative analysis with athlete guidance such as might appeal to the coach. In addition theoretical and physical models are presented that have direct, practical relevance.

It is our sincere hope that in representing the ideas and endeavours of a few outstanding individuals currently working at the leading edge of sports biomechanics research, this book will be welcomed as a source of interest and perhaps even of inspiration by those actively engaged in the field.

The editors

Med. Sport Sci., vol. 25, pp. 1–18 (Karger, Basel 1987)

Muscle Elasticity and Human Performance

Martyn R. Shorten[1]

Nike Sports Research Laboratory, Beaverton, Oreg., USA

Introduction

Elasticity in muscle fibres and tendons plays an important role in enhancing both the effectiveness and the efficiency of human performance. Neither muscle nor tendon behaves like a perfect spring, but both possess mechanical properties that can be described by relatively simple elastic models. Of particular importance is the ability of these tissues to store energy when deformed by a force and to recoil after being stretched.

Elastic behaviour is characterised by the relationship between the deformation of an elastic structure and the force applied to it. For an ideal linear spring, deformation x is a linear function of the force, F:

$$F = kx, \tag{1}$$

where $k = dF/dx$ is the stiffness of the spring. The inverse of stiffness, dx/dF, is called the compliance of the spring. When a force F compresses or stretches a spring, the work done by the force is stored as strain energy in the deformed structure. The amount of energy (E) stored in an ideal spring of stiffness k is equal to the work done (W) by the stretching force and can be determined by integration of equation (1):

$$\Delta E = W = \int F \cdot dx = \int kx \, . \, dx = \tfrac{1}{2}k \cdot x^2 \, . \tag{2}$$

Another important feature of elastic systems is the existence of a natural frequency of oscillation. When disturbed from its static equilibrium position,

[1] The author wishes to thank Dr. Gordon Valiant for his constructive comments and critical review of the manuscript which brought about many improvements. Les Cooper, Richard Durost and Bob Mueller also made a valuable contribution by providing technical support and their assistance is gratefully acknowledged.

Mass spring

Equation of motion: $M\ddot{x} + kx = Mg$
Natural frequency: $\omega_n^2 = k/M$

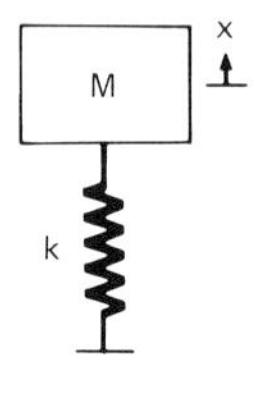

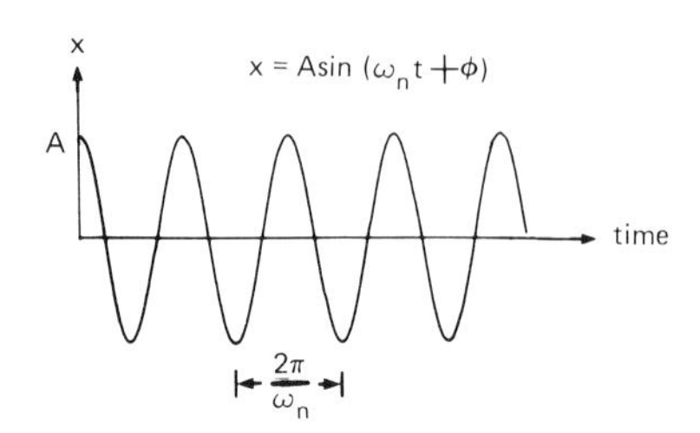

Damped mass spring

Equation of motion: $M\ddot{x} + c\dot{x} + kx = Mg$
Damping ratio: $\zeta = C/2M\omega_n$
Damped frequency: $\omega_d^2 = \omega_n^2\,(1-\zeta^2)$

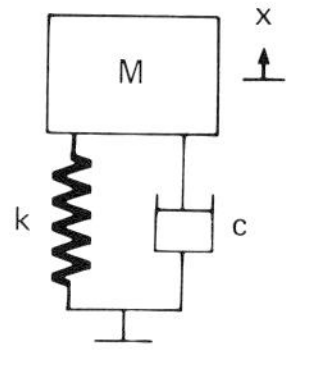

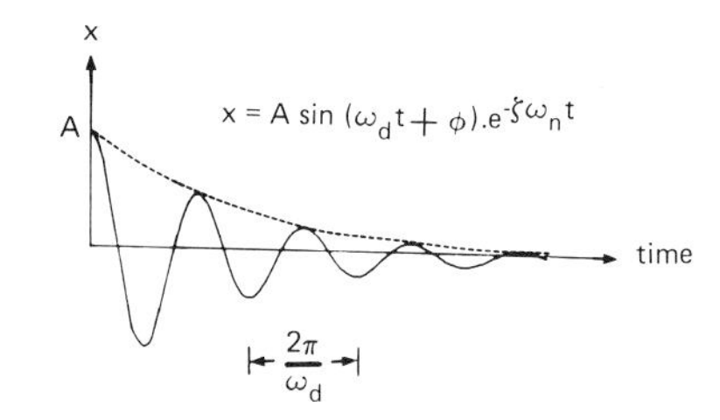

Fig. 1. Equations of motion and free oscillations of mass-spring systems *(a)* without damping and *(b)* with viscous damping of C $N \cdot m^{-2}$. For a more complete analysis, see Thompson [1981].

a mass-spring system (fig. 1a) vibrates at a characteristic natural frequency ω_n, which is determined by the mass m, and the spring stiffness k. When viscous damping is present (fig. 1b), the oscillation frequency is a function of the natural frequency and the damping; and the damped oscillations fade exponentially at a rate which also depends on the damping.

Elasticity in Muscle and Tendon

The stiffness characteristics of living tissues are variable, non-linear and subject to damping. Elastic and viscous elements must therefore be combined to describe their behaviour.

Classical models [Fenn and Marsh, 1935; Hill, 1938] describe the behaviour of muscle in terms of three functionally distinct components (fig. 2). The contractile component (CC) represents the force generating processes in the muscle, described by the characteristic force-velocity and force-length relationships [Hill, 1938; Wilkie, 1950; Gordon et al., 1966; Ralston et al.,

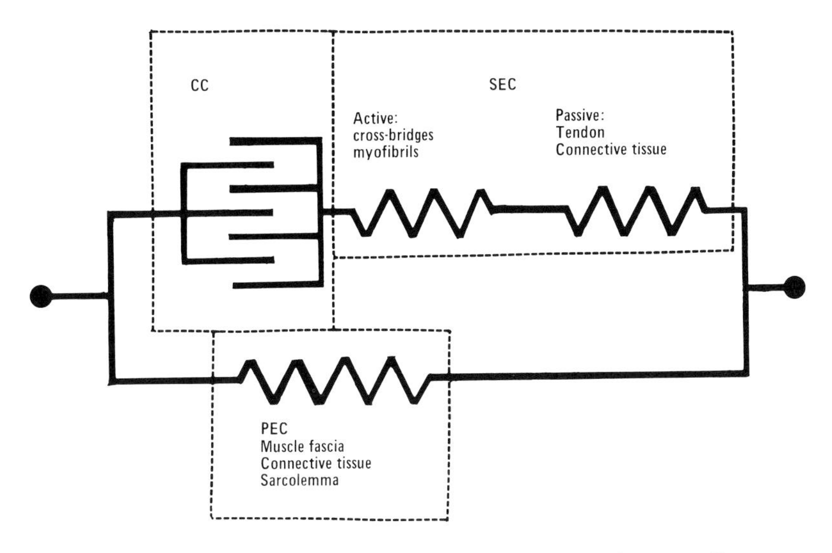

Fig. 2. A three-component mechanical model of muscle consisting of a contractile component (CC), a parallel elastic component (PEC) and a series elastic component (SEC) made up of active and passive elements.

1947]. The series elastic component (SEC) and parallel elastic component (PEC) represent groupings of anatomically distributed elastic structures according to their geometric relationship with the contractile component. Structures in series with the contractile component transmit the force of a muscular contraction. Structures in parallel are not brought under tension by contractile forces but carry the passive tension across a joint when the muscle is inactive.

A major portion of series elasticity resides in the tendon. Since tendon stiffness is largely independent of contractile component activity it can be considered a passive component of the SEC. In addition, some of the series elasticity lies within the contractile proteins which bear tension when the muscle contracts. Cross-bridge linkages and, to a lesser extent, the myofibrils themselves, are non-rigid and stretch slightly under load. Cross-bridge compliance is especially important, with a single cross-bridge able to stretch by 15–30 nm [Huxley and Simmons, 1971; Rack and Westbury, 1974].

Since both muscle force and the active component of stiffness depend upon the number of cross-bridges attached, the stiffness of the active part of the SEC increases with increasing muscle tension [Haugen, 1982; Sonnenblick, 1964; Morgan, 1977]. The rise in stiffness with increasing tension is generally linear and is independent of muscle length, stimulus rate and fatigue [Morgan, 1977; Goubel and Pertuzon, 1973; Hunter and Kearney, 1983].

With the model configuration shown in figure 2, tension is borne by the PEC alone when the CC is inactive. The muscle fasciae and sarcolemma are the main sites of parallel elasticity in the whole muscle but interaction between filaments and residual cross-bridge attachments may also contribute to resting tension. Resting joint stiffnesses are very low compared with the stiffnesses of muscle and tendon. Typical parallel angular stiffnesses (for the mid-range of joint motion) are 21Nm/rad for the human hip joint [Yoon and Mansour, 1982], 12 Nm/rad for the ankle [Seigler et al., 1984] and 1.3 Nm/rad for the elbow joint (Hayes and Hatze, 1977].

In vivo Measurements of Human Muscle Elasticity

Some researchers have adapted methods used in isolated animal muscle experiments, such as the quick-release method [Jewell and Wilkie, 1958], to measure the elasticity of human muscles [Goubel et al., 1971; Goubel and Pertuzon, 1973; Cnockaert et al., 1978]. Others have modelled the body or a limb as a mass-spring system and calculated the visco-elastic properties of the spring from the natural frequency of free oscillations [Cavagna, 1970] or from the frequency response of the system to applied vibration [Aruin et al., 1978; Hunter and Kearney, 1982, 1983; Kearney and Hunter 1982]. The problem of distinguishing between individual muscles is usually avoided by assuming that a single equivalent muscle acts to flex or extend a particular joint.

Figure 3 illustrates a method of determining the viscoelastic properties of the human one-joint ankle extensor (plantar flexor) muscles using a free vibration technique. Subjects sit with their forefeet on the edge of a force-plate and support a frame loaded with weights on the knees. With the knee joint flexed, it is assumed that the gastrocnemius muscles are too short to make a significant contribution and that the load is borne by the structures that do not cross the knee, principally the soleus, and the achilles tendon. A gentle downward push is enough to initiate damped oscillations of the mass-spring system which occur at frequencies of 3–6 Hz and are lightly damped (fig. 1b). The damped oscillations are recorded by the force-plate system and used in conjunction with the equation of motion of a damped mass-spring model (fig. 3b) to calculate the elastic stiffness of the muscle. Repeating the experiment at a range of loads enables the relationship between muscle stiffness and force to be dermined.

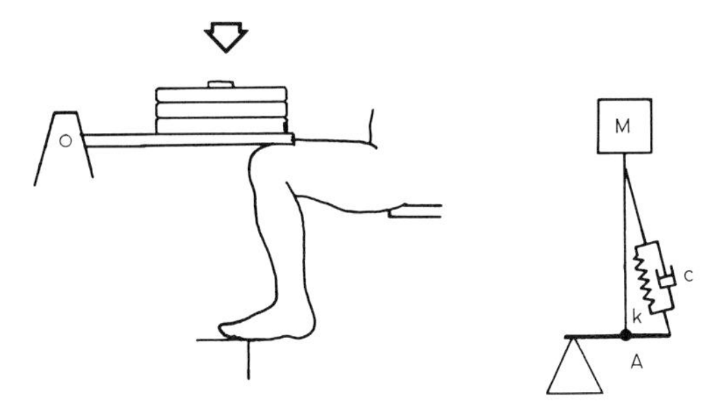

Fig. 3. a Experimental set-up for determining the viscoelastic characteristics of the human soleus muscles. *b* Damped mass spring model used to interpret free oscillations of the supported mass.

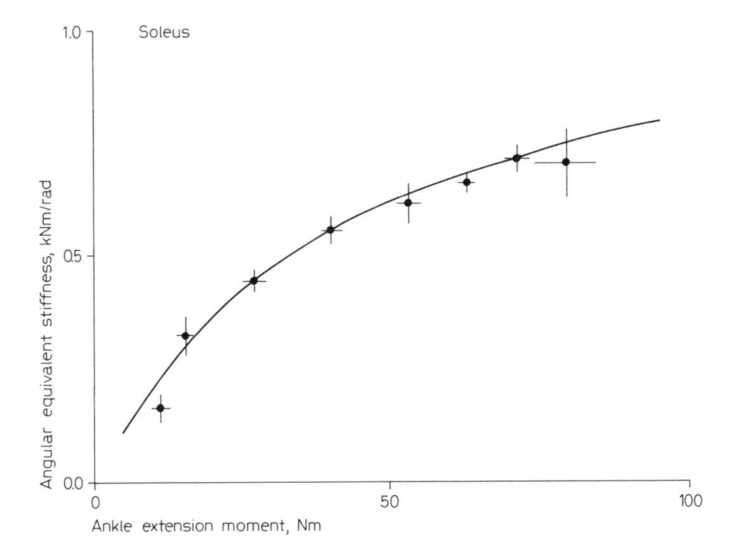

Fig. 4. Elastic stiffness of the human soleus muscle as a function of muscle load. Stiffness and load are expressed as angular equivalents with respect to the ankle joint (mean and SEM of 9 subjects). The line represents a model of the type shown in figure 5, with parameters k_T = 1091 Nm/rad, k_P = 14.7 Nm/rad and k_S = 27.0 rad^{-1}.

The stiffness-force function obtained during these experiments is shown in figure 4. The data are expressed in angular equivalent terms, with angular stiffness about the ankle joint plotted as a function of ankle extensor moment, and are the mean values for 9 subjects. The curvilinear increase in stiffness with increasing joint moment (muscle force) can be described by an

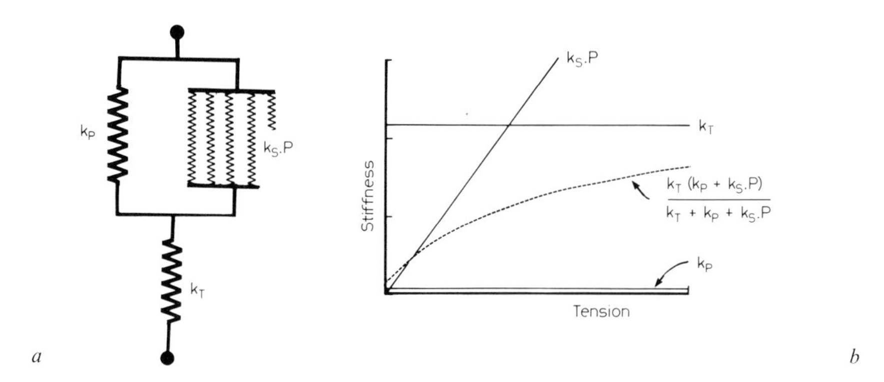

Fig. 5. a Interpretation of soleus muscle elasticity in terms of an elastic model incorporating a tendon stiffenss of k_T, a parallel stiffness of k_P and a series stiffness that is the product of tension P and a constant, k_S. *b* Contribution of individual components (solid lines) to the overall stiffness (dotted line) of the model.

elastic model based on the assumption that the stiffness of the active part of the muscle SEC is a linear function of muscle tension. Combining this active series component with a tendon and a parallel component results in the model configuration shown in figure 5, where k_P and k_T are the elastic stiffnesses of the parallel component and tendon respectively. The active series stiffness is described by $k_S.P$ where P is the contractile force and k_S is a constant. The overall stiffness of this model, λ, is given by:

$$\frac{1}{\lambda} = \frac{1}{k_T} + \frac{1}{k_P + k_S \cdot P}, \quad (3)$$

i.e.

$$\lambda = \frac{k_T \cdot (k_P + k_S \cdot P)}{k_T + k_P + k_S \cdot P}. \quad (4)$$

As tension (P) increases, the second term of equation (3) approaches zero and the overall stiffness approaches k_T, the stiffness of the tendon alone. Thus muscle tension appears to be the major determinant of stiffness at low loads, but as load increases so the stiffness tends towards that of the more compliant tendon.

Resonance

If a damped mass-spring system is subjected to a cyclic load, the amplitude response of the system depends on the relationship between the frequency of the cyclic force and the natural frequency of the system. The excursion amplitude of the damped mass-spring system is highest when the forcing frequency matches the natural frequency, and the system is said to resonate. At the resonant frequency, the magnitude of force required to achieve a given amplitude of movement in the system is at its lowest. Bach et al. [1983] have presented experimental evidence suggesting that the motor system can take advantage of resonance by selecting movement frequencies that coincide with the natural frequency of the mass-spring system formed by the body and the supporting musculature of the legs. Using a theoretical visco-elastic model of the arm, Denoth [1985] demonstrates that optimal performance of a simple throwing movement occurs when the effective frequency of contractile component activity matches the natural frequency of the SEC-mass system. It has also been shown that a running track may be tuned to the elastic characteristics of the human body, resulting in improved performance and fewer injuries [McMahon and Greene, 1979]. In theory, oscillatory movements performed at the natural frequency of muscle and tendon elastic components could maximise performance or minimise effort.

Elastic Energy Storage in Stretched Muscle and Tendon

When a concentric muscle contraction is preceded by a stretching eccentric phase, the force, power and work produced are greater than for a contraction without pre-stretch [Cavagna et al., 1968; Edman et al., 1978; Bosco and Komi, 1979]. Part of the improved performance following a stretch can be attributed to a reduction in electromechanical delay time (Norman and Komi, 1979; Cavanagh and Komi, 1979) and to an increase in muscle activation levels brought about by stretch reflexes; but muscle and tendon elasticity also play a role. After a stretch, the contraction velocity is augmented by the recoil velocity of the elastic components. Elasticity, therefore, has the effect of shifting the hyperbolic force-velocity relationship of the contractile component in favour of higher forces at a given velocity or higher velocities at a given force [Cavagna and Citterio, 1974; Edman et al., 1978; Bosco and Komi, 1979; Bosco, Viitasalo et al., 1982].

Stretching a muscle-tendon system also allows elastic storage and recovery of energy to occur. The amount of energy stored by stretching is given by the product of elastic stiffness and the square of the stretch distance (equation 2). For example, a human achilles tendon has an elastic stiffness of about 250 kN/m [Benedict et al., 1968]. A load of 250 N will therefore stretch the tendon by 1 mm (equation 1) and store 0.125 J of energy in the stretched tissue (equation 2). Similarly, it has been calculated that the achilles tendon is stretched by 18 mm during a typical running stride at moderate speed [Alexander and Bennet-Clark, 1977] and that 42 J of energy are stored. The nonlinear relationship between the amount of stretch and the energy stored (equation 2) results in proportionately greater energy storage at higher stretch extensions than at lower ones.

Not all of the work done in stretching a muscle is stored as strain energy. Some energy will be lost in overcoming the viscosity of the muscle. Also, not all of the strain energy can be usefully recovered. Since elastic energy storage in the active component of the SEC is dependant upon cross-bridge attachments, the energy can only be stored for as long as these attachments exist. When cross-bridges detach (through relaxation of the contractile component, stretching beyond the elastic limit or cross-bridge turnover) this component of strain energy is dissipated as heat. Some strain energy will be retained in the tendon and PEC while tension is maintained but even this is gradually dissipated. Aruin et al., [1979] have calculated that elastic energy storage in the human leg extensor muscles decays exponentially at a rate equivalent to a half-life of about 4 s. Since crossbridge turnover and stress-relaxation in visco-elastic tissues ensure that elastic energy storage is only temporary, elastic recoil will only make a substantial contribution to muscular performance if concentric contraction immediately follows the active muscle stretch.

With the same stretching force, a more compliant material will store more energy than a stiffer one. Since the compliant material undergoes greater deformation, the force will act over a greater distance and do more work.

A tendon is more compliant than active muscle fibres and as a result will usually be responsible for the greater proportion of any energy stored. The ultimate strain energy storage capacity of tendon is also greater than that of the muscle fibres. Using data for the stretch limit of cross-bridges and for the maximum isometric force in vertebrate skeletal muscle, Alexander and Bennet-Clark [1977] have calculated that the energy storage capacity of the cross-bridge linkages is in the range 2.4–4.7 J/kg muscle. In contrast, the strain energy storage capacity of tendon collagen is between 2000 and

9000 J/kg. As a result, the capacity for elastic energy storage will be greatest in muscle groups with long, compliant tendons.

The Stretch-Shorten Cycle

The importance of pre-stretch to human muscular performance is indicated by the frequency with which movements are preceded by a stretching countermovement. It is unusual for natural movements to begin from a static position. Jumping movements generally begin with a downward rather than an upward movement and throwing motions are naturally preceded by a countermovement in the opposite direction to that of the intended throw. Before initiating motion in the intended direction, the throwing, and jumping muscles must first overcome the momentum of the countermovement by contracting eccentrically while being forcibly stretched. This pattern of eccentric contraction followed without pause by a concentric contraction is known as the 'stretch-shorten cycle'.

Figure 6 illustrates the pattern of extensor muscle action about the knee joint in three basic human movements. When the muscles are producing a positive extensor moment, and the extension velocity is positive, the muscle is acting concentrically. When the extensor moment is positive but extension velocity is negative, the extensor muscles are applying a contractile force but are being forcibly extended so the contraction is eccentric. In rebound jumping (fig. 6a) there is a characteristic cycle of eccentric contraction followed by concentric contraction. The leg extensor muscles also go through a stretch-shorten cycle during the ground contact phase of running (fig. 6b). The eccentric stretch is less forceful than it is in rebound jumping but still substantial. In contrast, during cycling without toe-clips (fig. 6c), the leg extensor muscles act purely concentrically, and are not actively stretched.

Activities like running, jumping and throwing meet the criteria for useful elastic energy storage, i.e. they include a stretch-shorten cycle of forcible stretch (eccentric contraction) followed immediately by concentric contraction of the working muscle groups. In contrast, activities like swimming, rowing and cycling, even though they are repetitive in nature, do not normally include a significant active stretch of the major working muscle groups. Recovery is passive and muscles shorten in concentric contraction directly from a relaxed state. Without active stretch preceding the concentric contraction, performance relies more on contractile than elastic muscle properties.

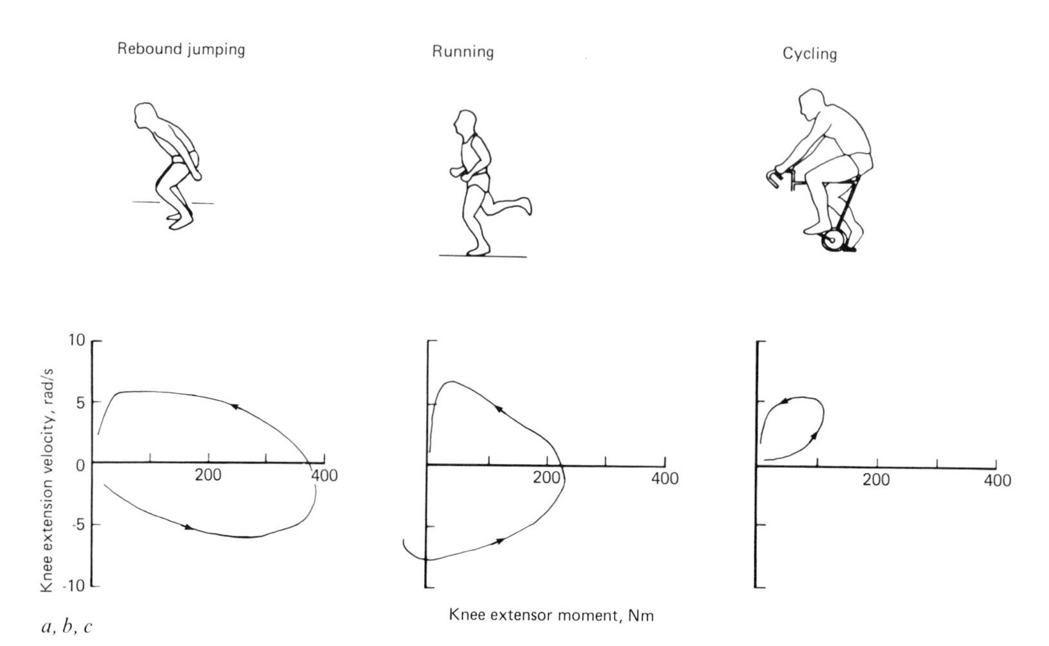

Fig. 6. Pattern of leg extensor muscle action during *(a)* rebound jumping, *(b)* support phase of running, and *(c)* cycling without toe-clips. The contractile force of the muscles is represented by the extensor moment about the knee joint, and the rate of contraction or stretch by the knee extension velocity. Both rebound jumping and running demonstrate a distinct stretch-shorten cycle of eccentric followed by concentric contraction but cycling does not.

Elasticity and Vertical Jump Performance

In human subjects, the most effective demonstrations of enhanced performance following a muscle stretch have been made by comparing vertical jumps performed with and without a preliminary countermovement. Bosco and Komi [1979] and Bosco, Ito et al. [1982], using a method first described by Asmussen and Bonde-Petersen [1974a], have made extensive comparisons of vertical jumps from a static starting position, jumps with a countermovement and jumps immediately following a drop to the ground from a height of 0.2–1.0 m. All the jumps were performed with maximal effort and all took place on a sensitive force plate. The force plate was used to measure the vertical ground reaction force and also allowed the jump height to be calculated. Jump heights are greater in the countermovement and drop jumps than in static jumps. The countermovement jump and drop jump allow a

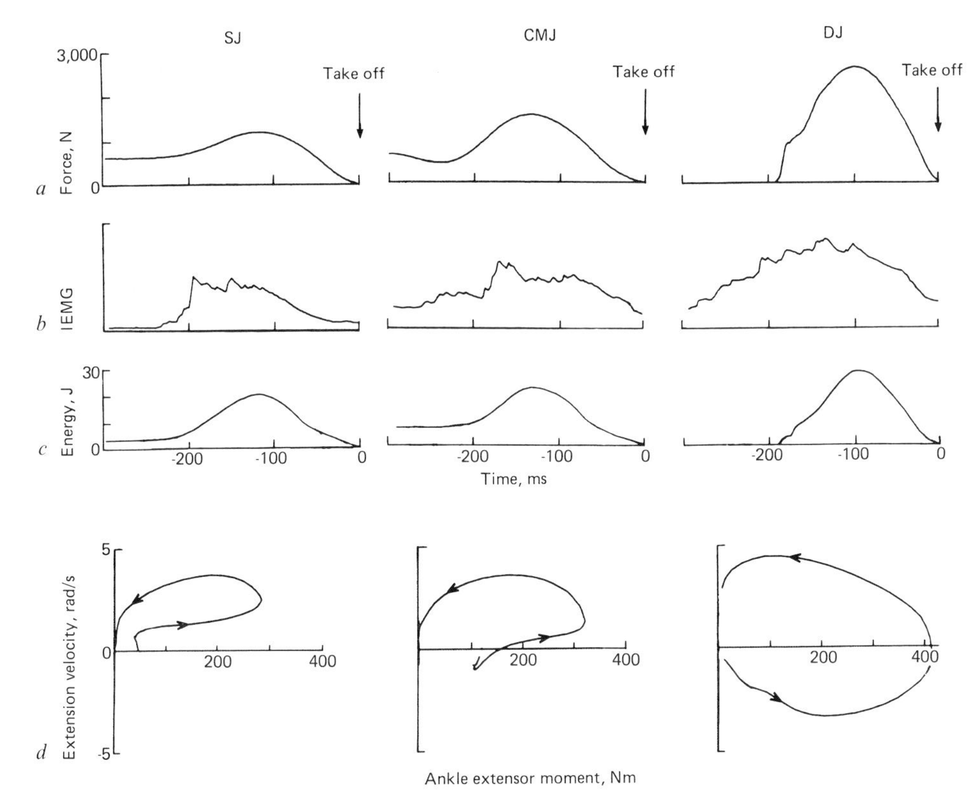

Fig. 7. Comparison of static jump (SJ) countermovement jump (CMJ) and drop jump (DJ) performed by plantar flexion of the ankles alone. *a* Vertical component of the ground reaction force. *b* Integrated electromyogram of the calf muscles. *c* Strain energy storage in these muscles. *d* Pattern of ankle plantar flexor muscle action showing different degrees of eccentric stretch under each condition.

forceful stretch of the elastic components of the leg extensor muscles before the concentric extension. Some strain energy can be stored in the stretched elastic components of the leg extensor muscles during the eccentric phase and recovered to enhance muscular performance during the concentric phase.

In the drop jump, jump height increases with increasing height of drop until an optimum 'stretch load', is reached, beyond which performance begins to deteriorate. Komi and Bosco [1978] measured the optimal drop height to be 62 cm in men and 50 cm in women, but the capacity to tolerate and make use of stretch varys among individuals. Optimum drop height in the drop jump is greater in volleyball players than in runners for example, suggesting an ability to tolerate higher stretch load. The difference between static jump and countermovement jump performances indicates how well athletes are able to make use of pre-stretch. In a study of 78 men from different sporting backgrounds, Bosco and Komi [1982] found that middle distance runners had a relatively small difference (5 cm) between static jump and countermovement jump performances while ski-jumpers had a much larger difference (8–11 cm).

Enhancement of jumping performance is significantly correlated with the speed of the pre-stretching movement and the brevity of the delay between eccentric stretch and concentric contraction [Bosco et al., 1981]. Subjects with more fast twitch fibres benefit more from a short, fast stretch of the leg extensor muscles while those with more slow twitch fibres are better able to make use of longer, slower stretches. Slow-twitch fibres have a longer cross-bridge cycle time and may therefore be able to make better use of long, slow stretches [Bosco, Tihanyi et al., 1982].

Contribution of Elastic Energy Storage to Enhanced Jumping Performance

While it is clear from experiments on isolated muscles without nerve connections that elastic energy storage contributes substantially to enhanced muscular performance after a stretch, other mechanisms may play a role in human subjects where the nervous connections are intact. Since EMG activity in the extensor muscles is much higher during a drop jump than in a static jump and can also be higher in a countermovement jump, the enhanced performance produced by a drop or countermovement must be attributed to a combination of elastic energy storage and additional, reflex-induced, myoelectric activity [Bosco, Viitasalo et al., 1982]. There is also evidence to suggest that reflexes act as a regulatory mechanism, possibly controlling muscle stiffness [Houk, 1979].

Using experimental data for the angular equivalent stiffness of the ankle extensor muscles it is possible to estimate the contribution of elastic energy storage in these muscles during simple jumping movements. Figure 7 shows some results from an experiment in which subjects performed simplified jumping movements from a force plate. The jumping movements were similar to the static jump, countermovement jump and drop jump described by Asmussen and Bonde-Petersen [1974a] and Komi and Bosco [1978] except that the subjects were trained to perform the jumps using only ankle extension (plantar flexion), while keeping the knees locked and the hands resting on the hips. Figure 7 includes an example of the relationship between ankle extension moment and ankle extension velocity for each jump, showing that during the static jump the ankle extension was purely concentric (extension velocity and extension moment are both positive). The countermovement jump included a brief eccentric contraction phase (extension moment is positive but extension velocity is negative) at the beginning of the jump. The drop jump condition is distinguished by the pronounced eccentric phase during the first part of contact with the ground. Some energy will be stored in the stretched elastic components whenever they are under tension. During the purely concentric static jump, 18.2 J of energy were stored in the stretched series elastic component but pre-jump eccentric contraction resulted in greater strain energy levels. The additional energy storage amounted to 2.6 J in the countermovement jump and 9.0 J in the drop jump. Relative to takeoff position, the subject raised his centre of mass by 4.1, 5.0 and 6.6 cm in static, countermovement and drop jumps, respectively. The subject had a mass of 86.0 kg so the improved jump height is equivalent to additional potential energy of 7.6 J in the countermovement jump and 21.1 J in the drop jump. This potential energy gain is greater than the energy conserved through elastic storage and recovery so only part of the observed stretch enhancement can be attributed to muscle elasticity. In this example, additional strain energy storage was 34% of the additional potential energy gained with a countermovement. In the drop jump, additional strain energy accounted for 43% of the additional potential energy gained. As figure 7 shows, the jumps with pre-stretch have higher extension moments at the onset of the concentric phase, higher activation (EMG) levels, higher peak moments and higher peak velocities as well as higher elastic energy storage; all of which could have contributed to the enhanced performance.

Elastic Energy Storage and Efficiency

In addition to making a muscle perform more effectively, the stretch-shorten cycle also results in more efficient performance. For example, subjects performing repeated knee bending exercises at the same rate use less oxygen if the upward movement immediately follows the downward movement [Thys et al., 1972]. A pause between the two phases of the movement results in a higher oxygen uptake, even though the net work done by the leg muscles is the same. The short coupling time between the eccentric phase and the concentric phase of the rebound movement allows some of the energy stored in the stretched muscle and tendon to contribute to the upward movement. Thus, some of the work done in raising the body is performed by recoil of the elastic components, reducing the requirement for muscular work and hence improving the efficiency of the movement. Similar results from other studies support the idea that the stretch and recoil of elastic components enhances the movement efficiency of men and other animals [Thys et al., 1975; Asmussen and Bonde-Petersen, 1974b; Bosco, Tihanyi et al., 1982; Cavagna et al., 1977]. Of particular interest are studies of hopping kangaroos which are endowed with long, compliant tendons and a remarkable potential for efficiency gains through elastic energy storage [Morgan et al., 1978; Alexander and Vernon, 1975; Dawson and Taylor, 1973].

Some indirect evidence for efficiency gains through elastic energy storage comes from attempts to measure the efficiency of complex human movements. The overall efficiency of the conversion of chemical energy to mechanical work in muscle is about 25% (i.e. 25% of the chemical energy released is converted to mechanical work and the rest is lost as heat). In activities such as cycling, where the muscles act concentrically without prestretch, the overall efficiency of the body is about the same as that of muscle. Cycling against a frictional load on a typical ergometer, for example, yields an efficiency ratio of up to 24% between work output and energy expenditure [Gaesser and Brooks, 1975]. In repetitive activities incorporating the stretch-shorten cycle, such as running and rebound jumping, measured efficiencies can be higher than in purely concentric exercise.

Most efficiency measurements made on running humans result in efficiencies well in excess of that of muscle being recorded, with values in the range 28–80% having been reported [Fenn, 1930; Cavagna and Kaneko, 1977; Fukunaga et al., 1980]. These high efficiencies appear to conflict with the conservation of energy principle. If work production by muscle has an efficiency of only 25%, the efficiency of the whole system cannot exceed this value unless there is some external energy input.

Transfer of potential and kinetic energy within and between body segments can account for over 70% of the total energy changes in a running stride [Shorten et al., 1981]. Transfer of energy makes muscle appear to be more efficient, because it allows more movement and more energy changes for the same amount of energy expended. When measuring the efficiency of human movement, failure to account for passive energy transfer leads to unusually high values being calculated.

Measurements of the efficiency of running, even with passive energy transfer accounted for, still result in high values. Some of the additional energy changes can be attributed to passive transfer of energy through elastic storage and recovery. Work-energy models of the human body that include passive energy transfer [Winter, 1979; Pierrynowski et al., 1980] can be extended to take account of elastic energy transformations [Williams and Cavanagh, 1983; Shorten, 1983]. In a study of ten athletes running on a treadmill at 4.5 ms^{-1} [Shorten, 1985], strain energy in the knee extensor muscles was found to increase as the extensor muscles contracted eccentrically during the first 100 ms of the support phase, reaching an average peak value of 66 J. The increase in strain energy coincides with a decrease in the total kinetic and potential energy of the body, suggesting that some energy transfer occurs. Without this energy storage and recovery, the net energy change during a running stride, and hence the muscular work requirement, would be 23% higher. Elastic energy storage can therefore increase the apparent efficiency of the body during the performance of activities with a stretch-shorten cycle.

Conclusion

In conclusion, muscle and tendon elasticity contribute to the effectiveness and efficiency of human performance through the mechanisms of force-enhancement, elastic energy storage and resonance. The influence of elasticity is greatest in activities which include the eccentric stretch of an active muscle group prior to a concentric contraction, i.e. a stretch-shorten cycle.

While the details of the interaction between muscle contraction, elastic properties and neural control mechanisms are not yet fully understood, it is clear that the muscle and tendon elasticity play an important role, complementing and enhancing muscle's contractile properties.

References

Alexander, R. McN.; Bennet-Clark, H.C.: Storage of elastic strain energy in muscles and other tissues. Nature, Lond. *265:* 114–117 (1977).

Alexander, R. McN.; Vernon, A.: The mechanics of hopping by kangaroos. J. Zool. *117:* 265–303 (1975).

Aruin, A.S.; Prilutski, B.I.; Raitsin, L.M.: Biomechanical properties of muscles and efficiency of movement. Hum. Physiol. *5:* 426–434 (1979).

Aruin, A.S.; Zatsiorski, V.M.; Panovko, G.Y.; Raitsin, L.M.: Equivalent biomechanical characteristics of the ankle muscles. Fiziol. Cheloveka *4:* 862–868 (1978).

Asmussen, E.; Bonde-Petersen, F.: Apparent efficiency and storage of elastic energy in human muscles during exercise. Acta physiol. scand. *92:* 537–545 (1974a).

Asmussen, E.; Bonde-Petersen, F.: Storage of elastic energy in skeletal muscles in man. Acta physiol. scand. *91:* 385–392 (1974b).

Bach, T.M.; Chapman, A.E.; Calvert, T.W.: Mechanical resonance of the human body during voluntary oscillations about the ankle joint. J. Biomech. *16:* 85–90 (1983).

Benedict, J.V.; Walker, L.B.; Harris, E.H.: Stress-strain characteristics and tensile strength of unenbalmed human tendon. J. Biomech. *1:* 53–63 (1968).

Bosco, C.; Ito, A.; Komi, P.V.; Luhtanen, P.; Rahkila, P.; Rusko, H.; Viitasalo, J.T.: Neuromuscular function and mechanical efficiency of human leg extensor muscles during jumping exercises. Acta physiol. scand. *114:* 543–550 (1982).

Bosco, C.; Komi, P.V.: Potentiation of the mechanical behaviour of the human skeletal muscle through prestretching. Acta physiol. scand. *106:* 467–472 (1979).

Bosco, C.; Komi P.V.: Muscle elasticity in athletes; in Komi, Exercise and sport biology, pp. 109–117 (Human Kinetics Publ., Champaign 1982).

Bosco, C.; Komi, P.V.; Ito, A.: Prestretch potentiation of human skeletal muscle during ballistic movement. Acta physiol. scand. *111:* 135–140 (1981).

Bosco, C.; Tihanyi, J.; Komi, P.V.; Fekete, G.; Apor, P.: Store and recoil of elastic energy in slow and fast types of skeletal muscles. Acta physiol. scand. *116:* 343–349 (1982).

Bosco, C.; Viitasalo, J.T.; Komi, P.V.; Luhtanen, P.: Combined effect of elastic energy storage and myoelectrical potentiation during stretch-shortening cycle exercise. Acta physiol. scand. *114:* 557–565 (1982).

Cavagna, G.A.: Elastic bounce of the body. J. appl. Physiol. *29:* 279–282 (1970).

Cavagna, G.A.; Citterio, G.: Effect of stretching on the elastic characteristics of the contractile component of the frog striated muscle. J. Physiol. *239:* 1–14 (1974).

Cavagna, G.A.; Dusman, B.; Margaria, R.: Positive work done by previously stretched muscle. J. appl. Physiol. *24:* 21–32 (1968).

Cavagna, G.A.; Heglund, N.C.; Taylor, C.R.: Mechanical work in terrestrial locomotion. Two basic mechanisms for minimising energy expenditure. Am. J. Physiol. *233:* 243–261 (1977).

Cavagna, G.A.; Kaneko, M.: Mechanical work and efficiency in level walking and running. J. Physiol. *268:* 467–481 (1977).

Cavanagh, P.R.; Komi, P.V.: Electromechanical delay in human skeletal muscle under concentric and eccentric conditions. Eur. J. appl. Physiol. *42:* 159–163 (1979).

Cnockaert, J.C.; Pertuzon, E.; Goubel, F.; Lestienne, F.: Series elastic component in normal human muscle; in Asmussen, Jorgensen, Biomechanics VIA, pp. 73–78 (University Park Press, Baltimore 1978).

Dawson, T.J.; Taylor, C.R.: Energetic cost of locomotion in kangaroos. Nature, Lond. *246:* 313–314 (1973).

Denoth, J.: Storage and utilisation of elastic energy in musculature; in Winter, Norman, Wells, Hayes, Patla, Biomechanics IXB, pp. 65–70 (Human Kinetics Publ., Champaign 1985).

Edman, K.A.P.; Elzinga, G.; Noble, M.I.M.: Enhancement of mechanical performance by stretch during tetanic contractions of vertebrate skeletal muscle fibres. J. Physiol. *281:* 139–155 (1978).

Fenn, W.O.: Frictional and kinetic factors in the work of sprint running. Am. J. Physiol. *92:* 583–611 (1930).

Fenn, W.O.; Marsh, B.S.: Muscular force at different speeds of shortening. J. Physiol. *85:* 277–297 (1935).
Fukunaga, T.; Matsuo, A.; Yuasa, K.; Fujimatsu, H.; Asahina, K.: Effect of running velocity on external mechanical power output. Ergonomics *23:* 123–126 (1980).
Gaesser, G.A.; Brooks, G.A.: Muscular efficiency during steady-state exercise. Effects of speed and work-rate. J. appl. Physiol. *38:* 1132–1139 (1975).
Gordon, A.M.; Huxley, A.F.; Julian, F.J.: The variation in isometric tension with sarcomere length in vertebrate muscle fibres. J. Physiol. *184:* 170–192 (1966).
Goubel, F.; Bouisset, S.; Lestienne, F.: Determination of muscular compliance in the course of movement; in Vredenbregt, Wartenweiler, Biomechanics II, pp. 154–158 (Karger, Basel 1971).
Goubel, F.; Pertuzon, E.: Evaluation de l'élasticité du muscle in situ par une methode de quick-release. Archs Physiol. Biochem. *81:* 697–707 (1973).
Haugen, P.: Short-range elasticity after tetanic stimulation in single muscle fibres of the frog. Acta physiol. scand. *114:* 487–495 (1982).
Hayes, K.C.; Hatze, H.: Passive viscoelastic properties of the structures spanning the human elbow joint. Eur. J. appl. Physiol. *37:* 265–274 (1977).
Hill, A.V.: The heat of shortening and the dynamic constants of muscle. Proc. R. Soc. Ser. B *126:* 136–195 (1938).
Houk, J.C.: Regulation of stiffness by skeletomotor reflexes. Ann. Rev. Physiol. *41:* 99–114 (1979).
Hunter, I.W.; Kearney, R.E.: Dynamics of human ankle stiffness. Variation with mean ankle torque. J. Biomech. *15:* 747–752 (1982).
Hunter, I.W.; Kearney, R.E.: Invariance of ankle dynamic stiffness during fatiguing muscle contractions. J. Biomech. *16:* 985–991 (1983).
Huxley, A.F.; Simmons, R.M.: Proposed mechanism of force generation in striated muscle. Nature, Lond. *233:* 533–538 (1971).
Jewell, B.R.; Wilkie, D.R.: An analysis of the mechanical components of frog's striated muscle. J. Physiol. *143:* 515–540 (1958).
Kearney, R.E.; Hunter, I.W.: Dynamics of human ankle stiffness. Variation with displacement amplitude. J. Biomech. *15:* 753–756 (1982).
Komi, P.V.; Bosco, C.: Utilisation of stored elastic energy in leg extensor muscles by men and women. Med. Sci. Sports *10:* 261–265 (1978).
McMahon, T.A.; Greene, P.R.: The influence of track compliance on running. J. Biomech. *12:* 893–904 (1979).
Morgan, D.L.: Separation of active and passive components of short-range stiffness of muscle. Am. J. Physiol. *232:* 45–49 (1977).
Morgan, D.L.; Proske, U.; Warren, D.: Measurements of muscle stiffness and the mechanism of elastic energy storage in hopping kangaroos. J. Physiol. *282:* 253–261 (1978).
Norman, R.W.; Komi, P.V.: Electromechanical delay in skeletal muscle under normal movement conditions. Acta physiol. scand. *106:* 241–248 (1979).
Pierrynowski, M.R.; Winter, D.A.; Norman, R.W.: Transfers of mechanical energy within the total body and mechanical efficiency during treadmill walking. Ergonomics *23:* 147–156 (1980).
Rack, P.M.H.; Westbury, D.R.: The short-range stiffness of active mammalian muscle and its effects on mechanical properties. J. Physiol. *240:* 331–350 (1974).
Ralston, H.J.; Inman, V.T.; Strait, L.A.; Shaffrath, M.D.: Mechanics of human isolated voluntary muscle. Am. J. Physiol. *151:* 612–620 (1947).

Siegler, S.; Moskowitz, G.D.; Freedman, W.: Passive and active components of the internal moment developed about the ankle joint during human ambulation. J. Biomech. *17:* 647–652 (1984).
Shorten, M.R.: Mechanical energy models and the efficiency of human movement. Int. J. Modell. Simulat. *3:* 15–19 (1983).
Shorten, M.R.: Mechanical energy changes and elastic energy storage during treadmill running; in Winter, Norman, Wells, Hayes, Patla, Biomechanics IXB, pp. 313–318 (Human Kinetics Publ., Champaign 1985).
Shorten, M.R.; Wootton, S.A.; Williams, C.: Mechanical energy changes and the oxygen cost of running. Engng. Med. *10:* 213–217 (1981).
Sonnenblick, E.: Series elastic and contractile elements in heart muscle: changes in muscle length. Am. J. Physiol. *207:* 1330–1338 (1964).
Thompson, W.T.: Theory of vibration with applications, 2nd ed. (George, Allen & Unwin, London, 1981).
Thys, H.; Faraggiana, T.; Margaria, R.: Utilization of muscle elasticity in exercise. J. appl. Physiol. *32:* 491–494 (1972).
Thys, H.; Cavagna, G.A.; Margaria, R.: The role played by elasticity in an exercise involving movements of small amplitude. Pflügers Arch. Eur. J. Physiol. *354:* 281–286 (1975).
Williams, K.R.; Cavanagh, P.R.: A model for the calculation of mechanical power during distance running. J. Biomech. *16:* 115–128 (1983).
Wilkie, D.R.: The relation between force and velocity in human muscle. J. Physiol. *110:* 249–280 (1950).
Winter, D.A.: A new definition of mechanical work done in human movement. J. appl. Physiol. *46:* 79–83 (1979).
Yoon, Y.S.; Mansour, J.M.: The passive elastic moment at the hip. J. Biomech. *15:* 905–910 (1982).

Dr. M.R. Shorten, Nike Sports Research Laboratory, 9000 S.W. Nimbus Drive, Beaverton, OR 97005 (USA)

Med. Sport Sci., vol. 25, pp. 19–33 (Karger, Basel 1987)

Basic and Applied Research in the Biomechanics of High Jumping[1]

Jesús Dapena

Biomechanics Laboratory, Indiana University, Bloomington, Ind., USA

Introduction

The standards of excellence achieved by an athlete are a function of two distinct but interacting components of training: fitness and technique. Not only must the physical condition of the individual be optimized for a specific event, but technique must be fine-tuned continuously to obtain maximum benefit from the level of fitness attained.

Over the years, athletes have discovered ways of modifying their skills to improve performance. Some of these modifications are the outcome of trial and error, others may be acquired by imitation of the most successful of their fellow competitors, but many are probably the result of largely unconscious processes of development and change. Thus, most athletes will tend not to be fully aware of the exact nature of diverse elements of their technique. However, for any given sports event we should expect to find some elements that are advantageous, and therefore common to all those highly skilled in that event. There are also likely to be elements peculiar to given individuals or special subsets of athletes. If it were possible to identify the advantageous technique elements for a sports event and to understand in detail the mechanisms through which they work to improve performance, it would then be possible to check which elements are not being used, or not being used

[1] This project was supported in part by grants from the US Olympic Committee and The Athletics Congress. The author wishes to thank M. Feltner, F. Buczek, C. S. Chung, R. Bahamonde, and I. Oren for their help in data collection and production of computer graphics.

appropriately, by the individual athlete. It is with the investigation of such issues that biomechanical research is most concerned.

Biomechanical research directed to the improvement of sports performance may be separated into two main forms: (a) basic research, which strives to achieve a better understanding of the mechanisms involved in a sports event and to identify advantageous technique elements, and (b) applied research, which attempts to identify and correct the technique deficiencies of individual athletes. The purpose of this paper is to explain a process through which the knowledge obtained from both forms of research can be combined to provide athletes with information that may help them to improve their techniques. The process was carried out in our laboratory during a long-term project focused on the Fosbury-flop style of high jumping, and it is discussed here in that context.

Basic Research

The basic research stage started with a critical examination of the available literature. Many of the conclusions of the classic paper on the straddle style [Dyatchkov, 1968] were also applicable to the Fosbury-flop. However, most of the published work concerned with the achievement of a clearer understanding of the Fosbury-flop [Ecker, 1971, 1976; Dyson, 1972; Labescat, 1972; Beulke, 1973; Kerssenbrock, 1975] usually had little or no quantitative information to support the theories proposed. There were also some quantitative studies on the Fosbury-flop [Adachi et al., 1973; Nigg et al., 1974], but they were based on two-dimensional analyses of motion, a major limitation for the study of the markedly three-dimensional motions of the Fosbury-flop.

In the next step of the basic research stage, several studies were carried out. Their immediate objective was not the improvement of the performances of the high jumpers involved, but the acquisition of quantitative information that would permit a better understanding of the Fosbury-flop technique [Dapena, 1980a, b]. The high jump studies were preceded by the development of specific methodology for three-dimensional film analysis of high jumping [Dapena, 1978a] and for the calculation of the angular momentum of the human body [Hay et al.,1977; Dapena, 1978b]. The results of these high jump studies permitted an evaluation of the previous theoretical literature, and led to a more comprehensive mechanical explanation of the Fosbury-flop technique [Dapena, 1980a–c], that provided a sound basis for further research.

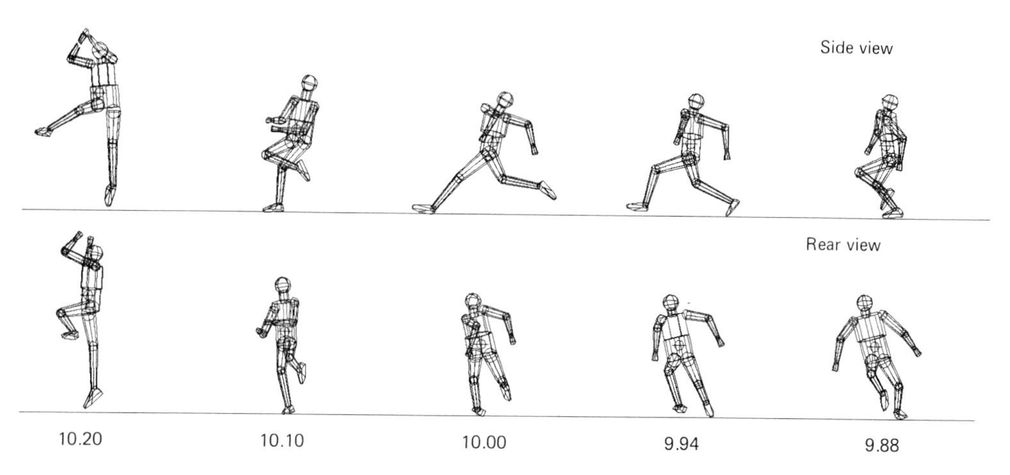

Fig. 1. Side and rear views of a high jumper during the final stride of the approach run and the takeoff phase. The numbers indicate time, in seconds. The time t = 10 s was arbitrarily assigned to the instant of the start of the takeoff phase.

Applied Research

From this point, two paths were open: further development of basic research [Dapena, 1979, 1981], and applied research [Dapena et al., 1986 a, b]. The applied research was part of the 'Elite Athlete Project', sponsored by the US Olympic Committee and The Athletics Congress. In this project, the high jumpers were filmed during official competitions. Computer programs were used to calculate three-dimensional body landmark coordinates throughout the last strides of the approach run, the takeoff, and the bar clearance [Dapena et al., 1982; Walton, 1981], to calculate other kinematic and kinetic parameters of the jumps [Dapena, 1978b, 1980a, b], and to produce computer plots that showed sequences of the jumps viewed from diverse directions (fig. 1, 2). These data were subsequently used to produce reports that explained to the athletes the advantages and disadvantages of their present techniques and gave recommendations on how to correct some of their technique problems.

The evaluation of the technique of each athlete started with an examination of the rotation over the bar and an analysis of the possible mechanisms

Fig. 2. Profile and overhead views of a high jumper during the bar clearance.

that might have led to any problems observed in it. In a second stage, the factors that may have had an influence on the maximum height reached by the center of mass were scrutinized. The present paper will concentrate on this second stage of the evaluation process, and the rationale behind some of the recommendations given to the athletes will be explained.

Approach Run Speed and Center of Mass Height

The most important part of a high jump is the takeoff phase. This phase is considered to start at the instant that the foot of the takeoff leg first makes contact with the ground. At that instant, the center of mass of the jumper usually has a large horizontal velocity (fig. 3a). In the 1984 US Olympic Trials, its average value was 7.4 m s^{-1} for the 12 male high jumpers analyzed, and 6.5 m s^{-1} for the 10 female high jumpers analyzed (all subsequent values cited in this section of the paper will refer to the same competition). During the takeoff phase, the ground pushes back on the athlete, reducing horizontal velocity (to 3.9 m s^{-1} in the men and 3.8 m s^{-1} in the women). This residual horizontal velocity gives the athlete the necessary horizontal displacement to reach the landing pit after the takeoff is completed.

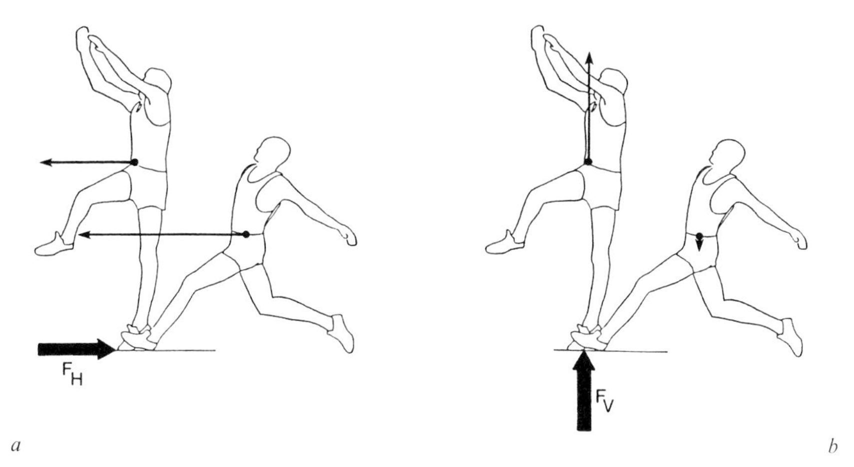

Fig. 3. Changes in the horizontal and vertical components of velocity of the center of mass during the takeoff phase, and forces responsible for them.

The vertical velocity at the start of the takeoff phase (fig. 3b) typically has a small negative value (men: -0.4 m s^{-1}; women: -0.3 m s^{-1}). During the takeoff phase, the athlete exerts a large downward force on the ground. The reaction to this force gives the athlete a large upward vertical velocity by the end of the takeoff phase (men: 4.2 m s^{-1}; women: 3.4 m s^{-1}). This vertical velocity component is the most important determining factor for the height of the parabolic path that follows the takeoff (fig. 4), and consequently for the result of the jump.

In order to maximize the vertical velocity at the end of the takeoff phase, the jumper needs to receive a large vertical impulse from the ground. That is, the product of vertical force and time of application should be as large as possible. A fast horizontal velocity at the end of the approach run may lead to an enhanced vertical force during the takeoff phase. This may happen in the following way: at the start of the takeoff phase, the takeoff leg is planted well ahead of the body (fig. 4, image 6). The momentum of the body makes the leg bend at the knee. The jumper tries to resist this bending, but the leg still flexes, stretching the knee extensor muscles (fig. 4, image 7). Eccentric muscle activity, the elastic component of muscle, and/or a stretch reflex mechanism may then act (in a way still not clearly understood) [Ozolin, 1973] to produce strong contractions of the extensor muscles of the takeoff leg,

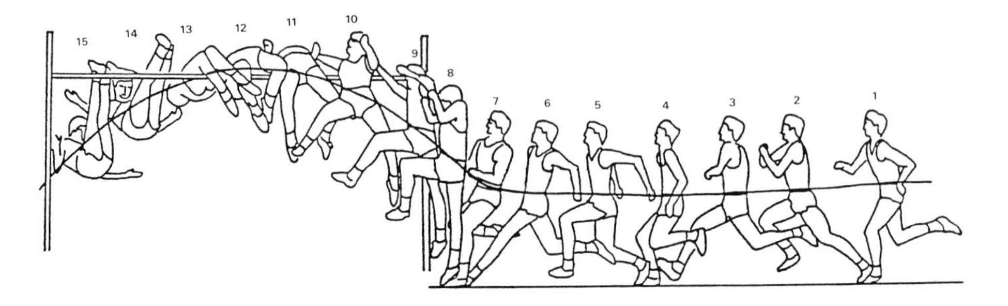

Fig. 4. The last two strides of the approach run, the takeoff phase, and the bar clearance of a typical Fosbury-flop. The curve indicates the trajectory of the center of mass.

leading to the exertion of a large vertical force before the jumper takes off from the ground (fig. 4, image 8).

The time during which vertical force is applied can be lengthened through an increase in the vertical range of motion covered by the center of mass during the takeoff phase ($\sim \Delta Z$ in fig. 5). For this, the center of mass has to be low at the start of the takeoff phase and high at the end of it. Most jumpers are reasonably high by the end of the takeoff phase, but it is difficult to be low at the start of the takeoff phase. This is because in that case the body has to be supported by a deeply-flexed non-takeoff leg during the penultimate stride (fig. 4, images 4–5), and this requires a very strong non-takeoff leg. It is also necessary to learn a rather unnatural pattern of movements for the last strides of the approach run. Consequently, a fast and low approach run can be achieved, but it requires a considerable amount of effort and training.

If an athlete learns how to run fast and low, there could be a new problem: the athlete may actually be too fast and too low. If the takeoff leg is not strong enough, it will be forced to flex excessively during the takeoff phase, and then it may not be able to make a forceful extension. In other words, the takeoff leg may "buckle" under the stress, resulting in an aborted jump. There probably is an optimum combination of approach run speed and center of mass height, and this optimum may not be the same for different athletes.

Figure 6 shows center of mass height at the end of the approach run (h_O, expressed as a percent of the standing height of each athlete) versus final speed of the approach run (V_H). The points show one jump by each of the 12 male high jumpers studied in the 1984 US Olympic Trials (the women had

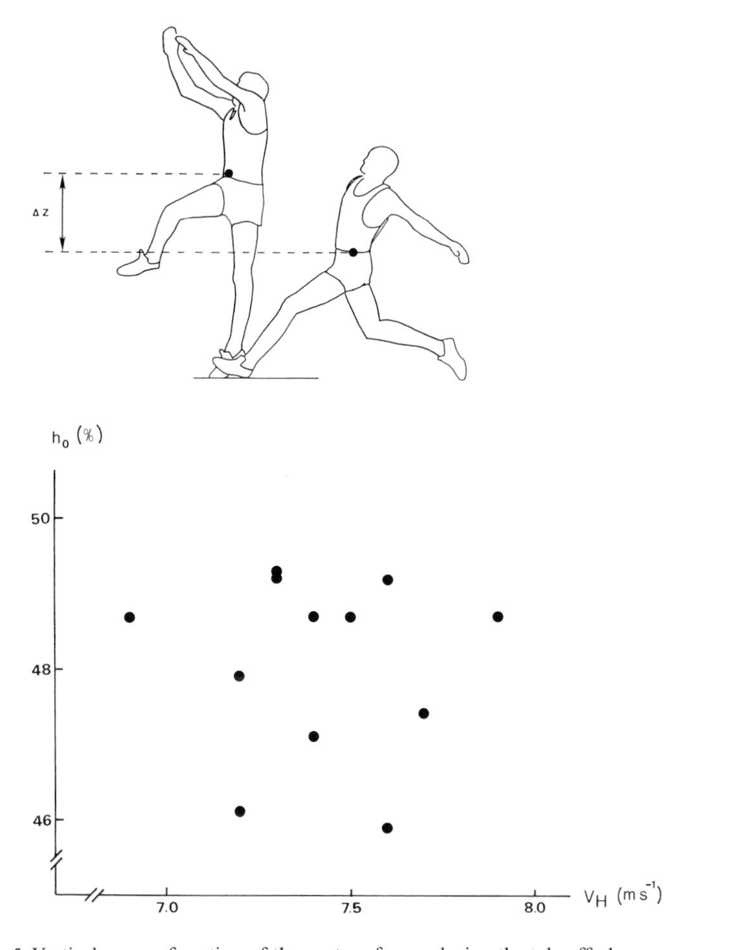

Fig. 5. Vertical range of motion of the center of mass during the takeoff phase.

Fig. 6. Plot of the height of the center of mass at the end of the approach run (h_O) versus the horizontal velocity of the center of mass at the end of the approach run (V_H). Each point represents one jump by one male high jumper.

smaller V_H values (5.9–6.9 m s^{-1}), and slightly larger h_O values (48.4–52.7%)).

Let us examine the probable effects of changes in the combination of h_O and V_H on performance. Figure 7 shows quadrants that designate four possible directions of change for one particular jumper. The null point represents the best analyzed jump of the athlete in the last competition. A change

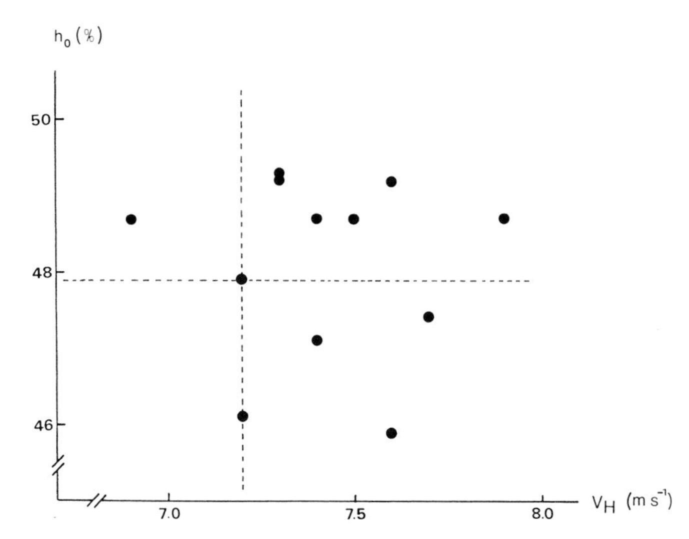

Fig. 7. Quadrants indicating options for changes in the height and horizontal velocity of the center of mass of a jumper (see text). The axis parameters are the same as in figure 6.

toward the upper left quadrant from the present position would imply slower velocity and higher center of mass. The athlete has probably tried such combinations before, because most young athletes start high jumping using slow and high approach runs: it is easier to run that way. Therefore, the athlete has probably tried before points in the upper left quadrant from his present position on the graph, but he is now at the lower right corner of it. This strongly suggests that the athlete jumps better in the present position: the takeoff leg is probably not buckling at that point. Consequently, it seems reasonable to assume that a change toward the upper left quadrant would result in a deterioration of performance.

If the position of the athlete changed toward the upper right quadrant, the desirable increase in velocity would be accompanied by an unwanted increase in center of mass height; if the position changed toward the lower left quadrant, the desirable decrease in center of mass height would be accompanied by an unwanted decrease in velocity. In these two cases it is not possible to say whether there would be an improvement or a deterioration of performance, but it seems reasonable to expect small changes in performance, given the combination of desirable and undesirable factors.

The lower right quadrant implies faster speed and lower center of mass. This should result in better jumps, unless the athlete is too fast and/or too low, in which case the takeoff leg will begin to buckle. If the athlete had experimented with jumps in this last quadrant and still decided to stay in its upper left corner, this would be a strong indication that the athlete already is at his/her optimum combination of speed and center of mass height, and that faster and lower approach runs would make the takeoff leg buckle. However, to be in the lower right quadrant a fast and low approach run is required, and it was indicated before that this is difficult to achieve, as considerable effort and training is necessary before this type of approach run can be mastered. Consequently, many athletes probably have never experimented with the lower right quadrant. Therefore, it is assumed for any athlete that the optimum combination of speed and center of mass height is, most likely, either at the present position of the athlete on the graph or somewhere in the lower right quadrant from it.

Jumpers are encouraged to learn a faster and lower approach run than their present one, and then to experiment jumping with it. If the athlete is able to jump higher than before, the new approach run should be retained; if the takeoff leg starts to buckle, the athlete should go back to the old technique. Most of the efforts for change are concentrated on the athletes in the upper left section of the graph shown in figure 6, because they are expected to be the most likely to benefit from faster and lower approach runs. The jumpers in the lower right section of the graph are more likely to be near their limits for buckling, and consequently, faster and lower approach runs are not stressed for them.

Techniques for Lowering the Center of Mass

The first strides of a high jump approach run are normal running strides. The center of mass is lowered only near the end, and this is achieved primarily through a combination of a lateral lean toward the center of the curve and flexion of the knee of the supporting leg. At the instant that the takeoff foot is planted on the ground to begin the takeoff phase, the center of mass should be comparatively low, and it should have a large horizontal velocity (see above). In a normal running stride, the center of mass has a downward vertical velocity component at the instant that the foot lands on the ground, but in the last stride of a high jump approach run it is important that the downward vertical velocity be minimized, in order not to waste effort in braking this downward motion during the takeoff phase. Consequently, the approach run of a high jumper should ideally lead to the following conditions at the start of the takeoff phase: large horizontal velocity, reasonably low center of mass, and minimal downward vertical velocity.

Figure 8 shows examples of three techniques used by high jumpers to lower the center of mass. The graphs correspond to three female jumpers with similar personal records (1.97–2.01 m). The abscissae show time, in seconds (the time t = 10.00 s was arbitrarily assigned to the instant of the start of the takeoff phase). The alternating shaded and unshaded bars at the bottom of each graph indicate ground support phases and airborne phases, respectively. The ordinates show height of the center of mass over the ground. To facilitate comparisons, the center of mass height values were divided by the standing height of each athlete and expressed as percent values.

The pattern shown in figure 8a was the most standard one: the center of mass was already low two strides before the start of the takeoff phase, and it was lowered slightly more in the last stride. At the end of the last support phase of the approach run (t = 9.93 s), the center of mass was low, and it had an upward vertical velocity of 0.4 m s^{-1}. When the takeoff foot finally made contact with the ground to start the takeoff phase (t = 10.00 s), the center of mass was low and it had only a moderate negative vertical velocity (–0.3 m s^{-1}). It would have been possible to reduce this negative vertical velocity somewhat by planting the takeoff foot earlier – this could be the mechanical basis for the usual advice of coaches to increase the tempo of the last two foot contacts of the approach run.

Figure 8b shows a very different technique. The athlete was very high two strides before the start of the takeoff phase. This running technique is more comfortable than that shown in figure 8a, but a normal takeoff phase requires a lower initial center of mass height. To achieve this, the athlete dropped very much during the last support phase of the approach run, and she had not quite stopped her downward motion by the end of this support phase (her vertical velocity at t = 9.95 s was –0.2 m s^{-1}). Her negative vertical velocity increased during the last airborne phase of the approach run, and by the start of the takeoff phase it had reached a very large magnitude (–0.7 m s^{-1} at t = 10.00 s). At the start of the takeoff phase, her center of mass was at approximately the same height as that of the athlete shown in figure 8a, but the large negative vertical velocity obliged her to make a much larger vertical impulse during the takeoff phase in order to jump approximately the same height.

Figure 8c shows an interesting third technique. In the middle of the last support phase of the approach run (t = 9.85 s) the center of mass of this athlete was lower than those of the other two athletes, but in the second half of the support phase the athlete lifted her center of mass considerably, and by the end of it (t = 9.95 s) she had a rather large upward vertical velocity (0.5 m s^{-1}). The subsequent airborne phase was very brief. By the beginning of the takeoff phase (t = 10.00 s), her center of mass was not any lower than those of the other two jumpers, but her vertical velocity was 0.0 m s^{-1}.

At this point, it is not possible to decide whether the athlete shown in figure 8c would have been better off maintaining a lower trajectory in the last stride, at the expense of a moderate negative vertical velocity at the start of the takeoff phase, or with her present technique, in which she sacrificed part of the previous lowering of her center of mass in order to avoid having any negative vertical velocity at the start of the takeoff phase. The inadequacies of the technique shown in figure 8b, however, seem clear. The high center of mass path in the penultimate stride made running more comfortable for the athlete, and therefore it facilitated the attainment of a fast approach run, but the penalty was high: a large negative velocity at the start of the takeoff phase. This athlete has been a very successful jumper, but her results should be expected to become still better if she adopted the techniques shown in figures 8a or 8c.

The advice to high jumpers is to learn how to run fast over a reasonably-flexed leg in the support phase immediately preceding the takeoff phase. After that, the athlete has two sensible options: (1) to maintain a reasonably low center of mass during the last stride at the expense of a moderate negative vertical velocity at the start of the takeoff phase (fig. 8a), or (2) to reduce the negative vertical velocity at the start of the takeoff phase at the expense of a slightly higher center of mass position (fig. 8c).

Action of the Arms and Leading Leg

The action of the arms and the leading leg during the takeoff phase of a high jump are very important for the outcome of the jump. As these limbs are accelerated upward during the takeoff phase, they exert by reaction a compressive force downward on the trunk. This force is transmitted through the takeoff leg to the ground. The increased vertical reaction force exerted by the ground on the athlete leads to a greater vertical velocity of the center of mass by the end of the takeoff phase, and consequently to a higher jump.

The effectiveness of the arm and leading leg actions can be evaluated in various ways. In the case of the arms, for example, the maximum vertical force or the total vertical impulse exerted on each arm could be calculated. However, these values are influenced by the motion of the trunk – for instance, if an athlete kept the arms completely inactive, they would still have the same vertical acceleration as the trunk. To avoid this problem, the verti-

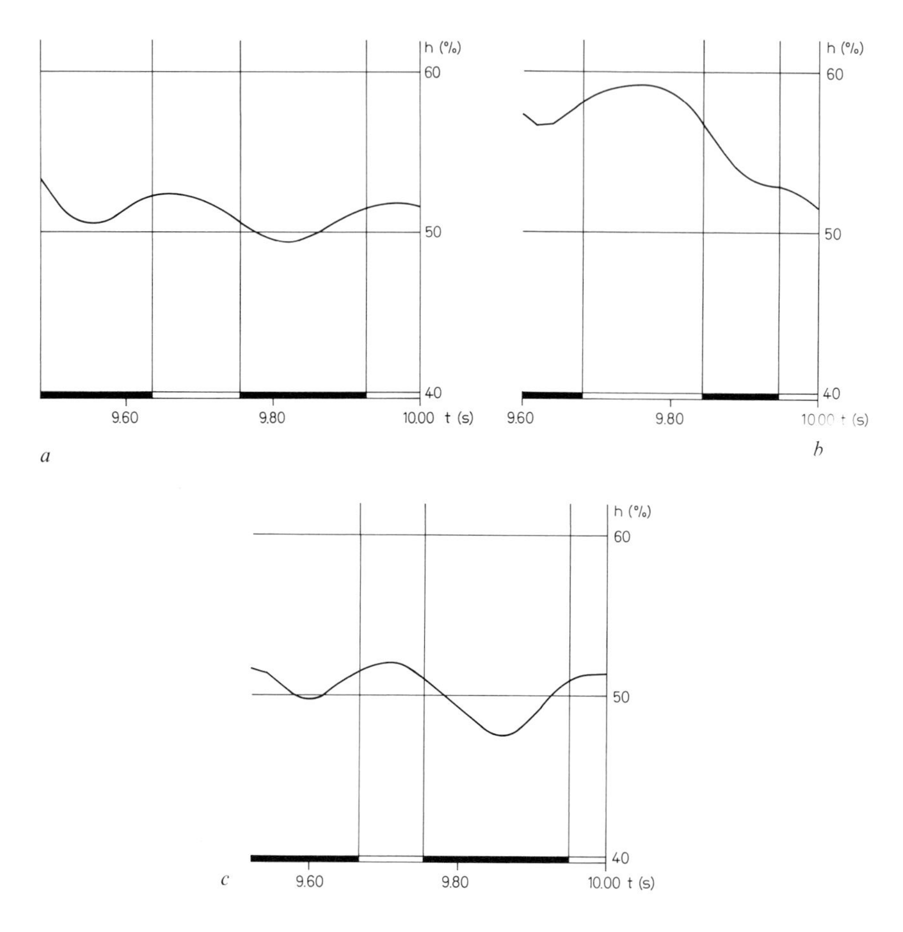

Fig. 8. Plots of center of mass height versus time for three female athletes in the final part of the approach run (see text). The shaded and blank bars indicate ground-support and airborne phases, respectively; t = 10.00 s is the instant of landing of the takeoff foot.

cal velocity values of the arms relative to the center of mass of the trunk were calculated. The difference between the minimum and maximum values of this relative velocity during the takeoff phase was calculated, and termed 'arm activeness'.

Figure 9 shows a plot of the activeness of the arm farthest from the bar (AAF), versus the activeness of the arm nearest to the bar (AAN), for a number of male and female elite high jumpers. The diagonal line indicates

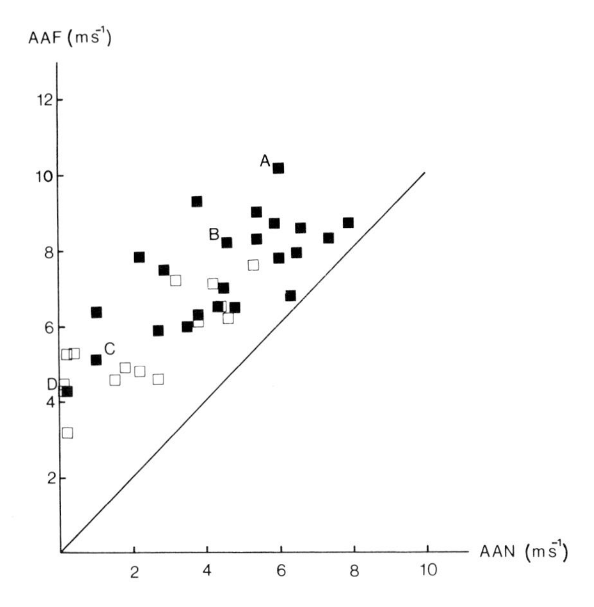

Fig. 9. Plot showing the activeness of the arm farthest from the bar (AAF) versus the activeness of the arm nearest to the bar (AAN). Male jumpers are represented by the solid symbols; female jumpers, by the open symbols. The letters A–D mark the four jumpers shown in figure 10.

the points for which both arms would have the same activeness. The arm farthest from the bar (i.e. the left arm for jumpers that take off from their left foot) was the most active in all cases.

Figure 10 shows sequences of the jumpers marked by the letters A–D in figure 9. Jumper A had the strongest arm actions, and jumper D the weakest. The sequences show that the jumpers with the best arm actions are those that have the arms farthest back at the start of the takeoff phase and the smallest amount of elbow flexion during the takeoff phase. It should be pointed out, however, that the evaluation of the arm actions of athletes always has to take into account other mechanical aspects of the jump. For instance, jumper C had a very fast approach run (7.9 m s^{-1}), and thus he may not have had enough time to prepare or execute a more vigorous arm action. Attempts to make substantial modifications in his arm action could lead to a decrease in the speed of the approach run, and this could be counterproductive. Consequently, major changes in the pattern of the arm actions are only advised for athletes that have slower than average approach runs.

Fig. 10. Sequences showing the last stride and the takeoff phase of the four jumpers designated with the letters A–D in figure 9.

Conclusions

In conclusion, it can be said that biomechanical investigations along the lines discussed above give real promise of helping athletes to improve their performance. However, full understanding has yet to be achieved, even in the case of the high jump, and, consequently, any recommendations made should be adopted with reservations. It is to be hoped that these reservations will ease as more information is made available to the athlete and the coach through the processes of basic research.

References

Adachi, N.; Asami, T.; Togari, H.; Kikuchi, T.; Sano, Y.: An analysis of movement in backward-roll style high jump. Proc. Dept. Phys. Ed. Univ. Tokyo; *7:* 69–75 (1973).

Beulke, H.: Der physikalische Sinn des bogenförmigen Anlaufs beim Fosbury-flop. Lehre Leichtathletik *51/52:* 1823–1828 (1973).

Dapena, J.: Three-dimensional cinematography with horizontally panning cameras. Sci. Motricite *3:* 3–15 (1978a).

Dapena, J.: A computational method for determining the angular momentum of a human body about three orthogonal axes passing through its center of gravity. J. Biomech. *11:* 251–256 (1978b).

Dapena, J.: A simulation method for predicting the effects of modifications in human airborne movements; Unpubl. doctoral diss.; University of Iowa (1979).

Dapena, J.: Mechanics of translation in the Fosbury-flop. Med. Sci. Sports Ex. *12:* 37–44 (1980a).

Dapena, J.: Mechanics of rotation in the Fosbury-flop. Med. Sci. Sports Ex. *12:* 45–53 (1980b).

Dapena, J.: The Fosbury-flop technique. Track Field Qu. Rev. *80:* 22–27 (1980c).

Dapena, J.: Simulation of modified human airborne movements. J. Biomech. *14:* 81–89 (1981).

Dapena, J.; Feltner, M.; Bahamonde, R.: Biomechanical analysis of high jump, No. 5 (Men). Report for Scientific Services Project; USOC/TAC (US Olympic Training Center, Colorado Springs 1986a).

Dapena, J.; Feltner, M.; Bahamonde, R.; Chung, C.S.: Biomechanical analysis of high jump, No. 4 (Women). Report for Scientific Services Project; USOC/TAC (US Olympic Training Center, Colorado Springs 1986b).

Dapena, J.; Harman, E. A.; Miller, J. A.: Three-dimensional cinematography with control object of unknown shape. J. Biomech. *15:* 11–19 (1982).

Dyatchkov, V.M.: The high jump. Track Technique *34:* 1059–1074 (1968).

Dyson, G.: The mechanics of athletics (University of London, London 1972).

Ecker, T.: Track and field dynamics (Tafnews, Los Altos 1971).

Ecker, T.: Track and field – technique through dynamics (Tafnews, Los Altos 1976).

Hay, J. G.; Wilson, B. D.; Dapena, J.; Woodworth, G. G.: A computational technique to determine the angular momentum of a human body. J. Biomech. *10:* 269–277 (1977).

Kerssenbrock, K.: Klicove prvky v technice zadoveho spusobu. I. Atletika *6:* 20 (1975).

Labescat, C.: An interpretation of the Fosbury technique; in Wilt, The jumps, pp. 19–23 (Tafnews, Los Altos 1972).

Nigg, B.; Waser, J.; Biber, T.: Hochsprung; in Nigg, Sprung, springen, Sprünge, pp. 75–104 (Eidg. Tech. Hochschule, Zürich 1974).

Ozolin, N.: The high jump takeoff mechanism. Track Technique *52:* 1668–1671 (1973).

Walton, J.S.: Close-range cine-photogrammetry: a generalized technique for quantifying gross human motion; Unpubl. doctoral diss.; Pennsylvania State University (1981).

Dr. Jesús Dapena, Department of Physical Education, HPER, Indiana University, Bloomington, IN 47405 (USA)

Med. Sport Sci., vol. 25, pp. 34–45 (Karger, Basel 1987)

Mechanical Power in Human Movement: Generation, Absorption and Transfer

David A. Winter

Department of Kinesiology, University of Waterloo, Canada

Most athletic movements are characterized either by precise and explosive bursts of energy generation and absorption, or by efficiently timed and regulated energy bursts designed for maximum efficiency. Biomechanists are fortunate that they can analyze specific performances by examining appropriate energy patterns at individual joints, within individual segments or within an entire limb. It must be realized that these analyses quantify the net effect of the neural control and metabolic systems. The final performance is a product of the efficiency of our metabolic system to convert from metabolic to mechanical energy and the efficiency of the neural control system to provide optimal control of that energy. In the assessment of pathological gait, there are many examples of both neurological and metabolic inefficiencies. Many neurological disorders (i.e. cerebral palsy) involve patients whose muscles and cardiorespiratory system are in excellent shape but whose neural control has lost most of its synergy (muscles are turning on at the wrong time, and not turning off at the right time). A biomechanical analysis of the motor patterns of such patients can pinpoint not only the muscle inefficiencies but also compensating patterns from muscles at other joints [Winter, 1985]. An extrapolation of these same techniques is probably the only way that biomechanics will have any major impact on improving athletic performance. The motor patterns that require emphasis are those that document the rate of energy generation, absorption and transfer (i.e. power). These power curves are the motor patterns that are the direct cause of the movement that is being observed. It is the flow of energy in or out of muscles that causes the detailed energy changes in each segment, and, in turn, the total body energy changes.

Review of Literature and Theory on Power and Energy

The vast majority of human movement research has focussed on walking and, as such, many researchers have commented on the tendency of the human body to conserve or optimize energy consumption [Bresler and Berry, 1951; Inman, 1966; Asmussen and Bonde-Petersen, 1974]. Such rationale has spurred theoretical simulations which have assumed some form of optimization. Unfortunately, these simulations [Crowninshield, 1978; Hardt, 1978; Seireg and Arvikar, 1975; Patriarco et al., 1981] have not been very successful in yielding valid predictions of motor patterns. Poor predictions were likely due to inadequacies in the simplified biomechanical model and the assumption that we walk with a single optimization criterion which controls our entire gait pattern.

What gives us some insight as to how efficiencies are achieved are analyses of the actual movement. Fenn [1929] was probably the first to attempt an accounting of conversion of metabolic to mechanical energy. Unfortunately, his analysis of sprinters ignored important energy exchanges within segments and transfers between adjacent segments. His power calculations were predictably high; the sprinter's average power was three horsepower! Elftman [1939a, b] in his classic studies introduced calculations of the rate of change of energy of the legs, the rate of energy transfer between segments across the joints and the rate at which muscles do work. Bresler and Berry [1951] reported potential, kinetic and total energies of the thigh, leg and foot during walking. Such a summation recognized energy exchanges within each segment. Force plate data were used by Cavagna and his coworkers Margaria, Komarek, Mazzoleni and Kaneko [1966, 1971, 1976] to calculate the energy of the total body centre of mass. There was one erroneous major assumption inherent in their calculation technique: that the energy of the body centre of mass represented the total energy of the body at any instant in time. Reciprocal movements of the arms and legs will not cause changes in the body's centre of mass, yet the energy requirements to swing the legs are not insignificant. Smith [1975] has pointed out that the body centre of mass is a weighted vector summation of all segment masses, whereas energies add as scalars. A segment-by-segment approach to correct for this error was introduced by Ralston and Lukin [1969], Winter [1976a, 1979] and some refinements on this approach have been offered by Williams and Cavanagh [1983]. Thus, it is mandatory to use techniques that recognize conservation of energy within each body segment and transfers of energy between adjacent segments. Such an approach requires analyses that calculate powers at each joint and within each segment.

As indicated earlier, Elftman [1939a, b] introduced some of the applicable techniques but, unfortunately, did not document his analyses with complete equations. Quanbury et al. [1975] documented the theoretical model of power generation, absorption and transfer by muscles, and passive transfers across joints, and presented results for the leg/foot segment during swing. Winter et al. [1976b] extended this to a three-segment swinging limb and a complete analysis during stance and swing was reported by Robertson and Winter [1980]. The theory behind the power analyses has been detailed in these previous papers and is summarized here. Assume we have two adjacent segments (fig. 1) rotating at angular velocities ω_1 and ω_2 with a moment of force, M, due to muscular activity.

The power generated by the muscles is:

$$P_m = M(\omega_2 - \omega_1) \text{ watts.} \tag{1}$$

Breaking this equation into two parts gives:

$$P_{m1} = M\omega_1 \tag{2}$$

and

$$P_{m2} = M\omega_2 \tag{3}$$

P_{m1} and P_{m2} are the rates of energy generation/absorption into or out of segments 1 and 2, respectively. These equations are written with the convention that counterclockwise moments of force and angular velocities are positive. Suppose, in figure 1, that we have a concentric contraction (i.e. $\omega_2 > \omega_1$). P_{m2} is positive indicating a rate of energy flow into segment 2. P_{m1} is negative because M is negative (the moment of force is clockwise on segment 1) and ω_1 is positive. This means that there are P_{m1} watts leaving segment 1. But the net power generated is $M(\omega_2 - \omega_1)$ which is less than the energy rate into segment 2. What is happening is that there is also an energy transfer from segment 1 to segment 2 which is equal to P_{m1}. Thus, the sum of the generation rate and transfer rate equals the rate of energy flow into segment 2. The rule of thumb to remember is that if two adjacent segments are rotating in the *same* direction (i.e. their angular velocities have the same polarity), there will be an energy transfer between the segments via the muscles and the magnitude of the transfer will be equal to lesser of P_{m1} or P_{m2}.

The second and passive mechanism for transfer of energy between segments is through the joint itself and is due to translational reaction forces

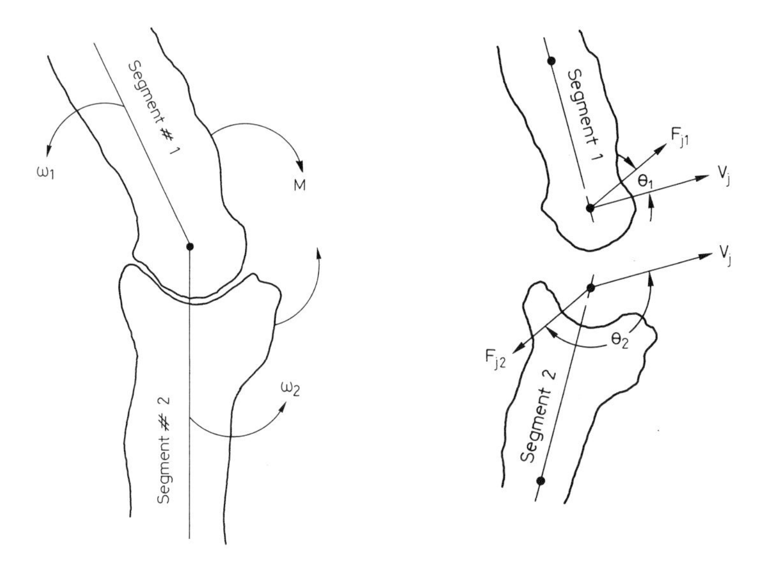

Fig. 1. Moment of force, M, and angular velocities, ω_1 and ω_2 associated with the mechanical power generation, absorption and transfer by the muscles.

Fig. 2. Force-velocity vectors at a joint centre. The dot product of these two vectors yields the rate of energy transfer across the joint centre.

acting at the joint centre. Figure 2 shows the reaction forces, F_{j1} and F_{j2}, acting on each segment, as the joint travels with a velocity, V_j.

The power into segment 1 from segment 2, P_{j1} is given by the dot product of the force and velocity vectors.

$$P_{j1} = F_{j1} \cdot V_{j*} = F_{j1}V_j \cos \theta_1. \quad (4)$$

Similarly, the power into segment 2, P_{j2}, is:

$$P_{j2} = F_{j2} \cdot V_j = F_{j2}V_j \cos \theta_2. \quad (5)$$

Note that P_{j1} is positive and P_{j2} is negative and that $P_{j1} = -P_{j2}$. In the situation depicted in figure 2, the fact that P_{j1} is positive means that energy is entering segment 1 from segment 2.

A summary of all possible combinations of muscle powers is reproduced in table I [Robertson and Winter, 1980].

Theoretically, by the Law of Conversation of Energy, we should be able to account for all energy flows into and out of any segment, and normally, for a given segment there are four possible routes: through the joint centres

Table I. All possible combinations of mechanical power generation, absorption and transfer by muscles crossing a joint, [reproduced with permission from Robertson and Winter, 1980]

Description of movement	Type of contraction	Directions of segmental ang. velocities	Muscle function	Amount, type and direction of power
Both segments rotating in opposite directions (a) joint angle decreasing	Concentric	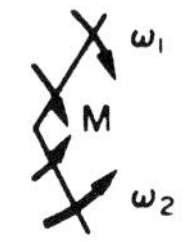 ω_1 M ω_2	Mechanical energy generation	$M\omega_1$ generated to segment 1. $M\omega_2$ generated to segment 2.
(b) joint angle increasing	Eccentric	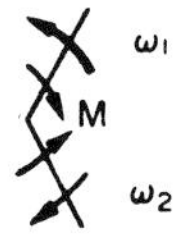 ω_1 M ω_2	Mechanical energy absorption	$M\omega_1$ absorbed from segment 1. $M\omega_2$ absorbed from segment 2.
Both segments rotating in same direction (a) joint angle decreasing (e.g. $\omega_1 > \omega_2$)	Concentric	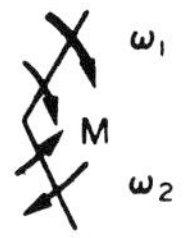 ω_1 M ω_2	Mechanical energy generation and transfer	$M(\omega_1 - \omega_2)$ generated to segment 1. $M\omega_2$ transferred to segment 1 from 2.
(b) joint angle increasing (e.g. $\omega_2 > \omega_1$)	Eccentric	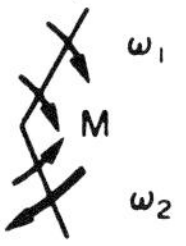 ω_1 M ω_2	Mechanical energy absorption and transfer	$M(\omega_2 - \omega_1)$ absorbed from segment 2. $M\omega_1$ transferred to segment 1 from 2.
(c) joint angle constant ($\omega_1 = \omega_2$)	Isometric (dynamic)	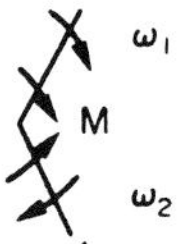 ω_1 M ω_2	Mechanical energy transfer	$M\omega_2$ transferred from segment 2 to 1.
One segment fixed (e.g. segment 1.) (a) joint angle decreasing ($\omega_1 = 0$, $\omega_2 > 0$)	Concentric	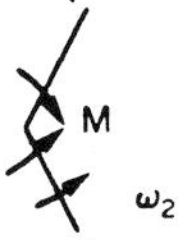 M ω_2	Mechanical energy generation	$M\omega_2$ generated to segment 2.
(b) joint angle increasing ($\omega_1 = 0$, $\omega_2 > 0$)	Eccentric	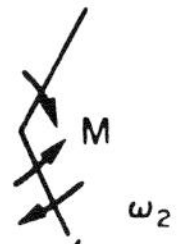 M ω_2	Mechanical energy absorption	$M\omega_2$ absorbed from segment 2.
(c) joint angle constant ($\omega_1 = \omega_2 = 0$)	Isometric (static)	 M	No mechanical energy function	Zero.

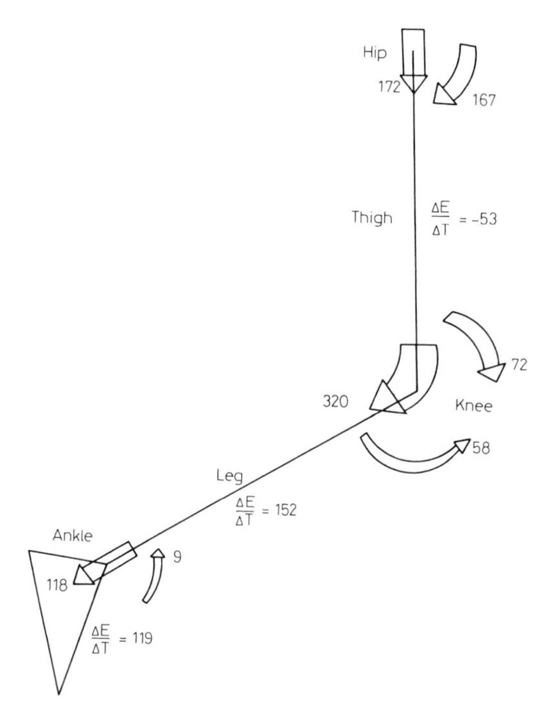

Fig. 3. Muscle and joint powers during early swing in the lower limb of a jogger. Arrows through the joint centre are the joint powers and the arrows around the joint are the muscle powers. The rate of change of energy of each segment, $\Delta E/\Delta T$, is calculated from the slope of the segment energy curve. See text for details of power balances. The joint powers show a general passive transfer of energy from the pelvis to the foot. All units in watts.

at the proximal and distal ends, and via muscles at the proximal and distal end. The sum of these four powers should equal the rate of change of energy of each segment. This power balance has been shown to be almost perfect experimentally during swing phase of walking [Robertson and Winter, 1980] and running [Winter, 1982]. Errors in the balance equation increase during stance and this was attributed to the high joint reaction forces combined with errors in the low velocities of the joints as a result of skin movement of the joint markers, or because the marker position did not reflect a perfect hinge joint.

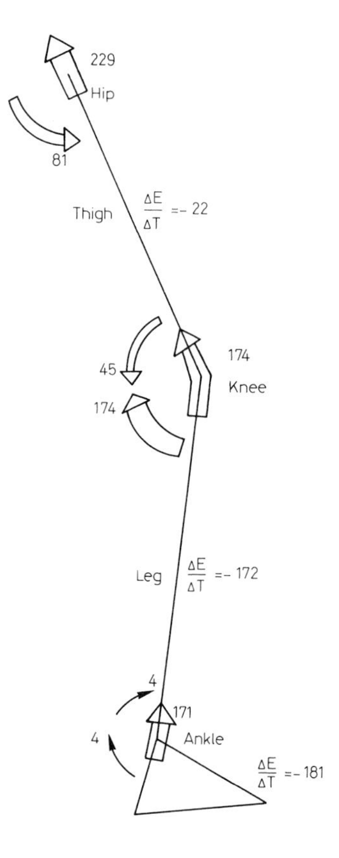

Fig. 4. The joint and muscle powers for the same runner as seen in figure 3 late in swing (approximately one-half stride later). A reversal of joint powers now occurs as energy passively flows from the foot to the pelvis, indicating a significant energy exchange with the contralateral limb. The power balance equation shows that the rate of energy change of each segment can be accounted for by an algebraic summation of the joint and muscle powers at proximal and distal ends. All units in watts.

Example Analyses of Running

Figures 3 and 4 are reproduced from analyses of a runner to show the situation at key phases during swing. Figure 3 summarizes the powers shortly after toe-off as the lower limb is being pulled upwards and forwards.

The arrows through the joint centre show the direction and magnitude of the intersegment transfers (equations 4 and 5) and the arrows around the joint show the contribution of the muscles (equations 2 and 3). Here, the passive energy rate through the hip joint was 172 W from the pelvis to the thigh, with an additional active contribution of 167 W from the hip flexors. The passive transfer across the knee joint to the leg was 320 W while the knee extensors were eccentrically contracting absorbing energy at a rate of 72 W from the thigh and 22 W from the leg. The ankle muscles had negligible activity, however, the foot is receiving energy passively through the ankle joint at a rate of 118 W. A power/rate-of-energy-change balance can be done, and via a separate calculation the rate of change of energy of the thigh was −53 W. The individual segment energies were calculated as described in Winter et al. [1976a] and the rate of change of energy was the slope of the total energy curve of the thigh. The four joint and muscle powers sum to 172 + 167 −320 −72 = −53 W, which is a perfect balance. It can be generally observed that the major mechanism to add energy to the leg and foot during early swing was via energy transfers through the joint centres. The hip flexor muscles were the only muscles actively involved in generating energy for the limb at this time but their contribution was less than one-half of the energy rates at the proximal end of the thigh.

A look at figure 4 for the lower limb late in swing shows a reverse flow of energy to be taking place. Energy leaves the foot, leg and thigh and most of it is conserved as it is transferred across the hip joint to the pelvis. There is a major absorption (174 + 45 W) by the knee flexors (mainly hamstrings) and a moderate generation (81 W) as the hip extensors actively extend the thigh.

When we consider figures 3 and 4 together, we are observing simultaneous events when one leg is starting its swing and the contralateral limb is decelerating prior to contact. Energy is seen to be flowing up one limb across the pelvis and down the opposite limb. Certainly, this transfer is a major mechanism which is responsible for efficient running. The question that remains to be answered is whether such a mechanism is optimized in successful athletes. Only a simultaneous analysis of both limbs of elite and non-elite athletes would answer the question. Alternatively, an analysis of the same athlete in rested and fatigued states would answer questions as to what aspect of energy generation, absorption or transfer starts to fail first. An even bigger challenge remains once an inefficiency has been diagnosed, because every movement must have some inefficiencies, and there is no information available as to what is optimal.

At the same time as recognizing these energy transfers across the joint centres and through the muscles, biomechanics researchers must also focus

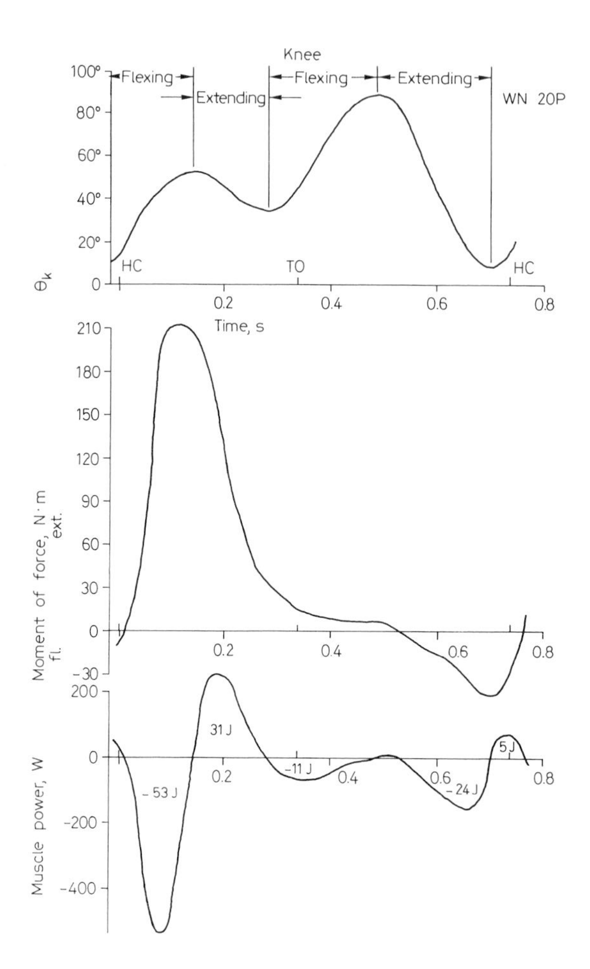

Fig. 5. Knee angle, moment and power during jogging. Knee angle plots shows that knee flexes from 5° to 45° during early stance, extends to 35° in mid-stance and flexes prior to toe-off (TO) reaching 85° in mid swing and extending back to 5° prior to heel contact (HC). Knee moment is extensor for all stance dropping to a low value early in swing, and changing to flexor for the latter half of swing. The power pattern shows a major absorption of energy (53 J) during weight acceptance, a modest generation (31 J) as the knee extends, a small absorption (11 J) to control amount of knee flexion during push-off and early swing, a major energy absorption (24 J) by hamstrings to decelerate leg and foot, and a small generation (5 J) by the same muscles to reverse the knee velocity and slow the leg's velocity to near-zero prior to heel contact.

on the patterns of energy generation and absorption by the muscles. Equation 1 calculates these muscle powers. Positive power takes place when the angular velocity has the same polarity as the moment of force, as in a concentric contraction. Negative power occurs when the angular velocity has the opposite polarity to the moment of force (eccentric contraction). Patterns for walking [Winter, 1983a] and running [Winter, 1983b] have been established. A typical power analysis is presented in figure 5, where it can be seen that each joint has distinct bursts of generation and absorption. Information from such analyses is significant in that it points to potentially important errors in training and conditioning. The plantar flexors are seen to be the most important energy generators and the quadriceps are primarily energy absorbers. Thus, isokinetic or other exercise machines may not be capable of loading the quadriceps eccentrically and are not likely to be able to load the plantar flexors at high loads over the normal range of movement and velocity. In a similar manner, power analyses have exposed certain 'myths', regarding race walking [White and Winter, 1985].

Chapman and Caldwell [1983] have documented the total energy flows across the joints in the swinging limb during treadmill running. They found the magnitudes of some of the peak powers to be thirty times higher than that seen during swing phase of walking. These peaks occur during much shorter bursts of time and, therefore, may represent only a tenfold increase of net energy generation, absorption, or transfer. The increases for swing during running are more drastic than they are for stance indicating that ballistic power becomes increasingly important in running compared with the powers associated with weight acceptance and push-off. In slow running, the push-off power [Winter, 1983b] still dominates and can reach peaks of 1500 W. However, it remains to be documented as to the power profiles in sprint running, but it would be safe to extrapolate and predict that the powers associated with swing would dominate those associated with weight acceptance and push-off.

Conclusions and Future Directions

At this time, the energetics of a total movement can be determined through these segment-by-segment and joint-by-joint analyses. The accuracy of such analyses are limited by the biomechanical model, which assumes hinge joints, constant length segments with constant masses, mass centres and moments of inertia. Also, the moment-of-force analyses yields the net algebraic summation of all muscle activity at each joint and is represented as a single torque motor. Thus, the level of co-contraction is not known until a partitioning of the individual muscle forces is known.

Even if a complete power analysis at the muscle level were available, researchers and coaches would be faced with indecision because of a lack of knowledge as to what is optimal. Theoretically, they could postulate what too many researchers have done in the past, that co-contractions are inefficient and must be eliminated. However, athletic movement would then become unsafe and be impossible to execute. How much co-contraction is necessary for joint stabilization is not known. Similarly, it could be postulated that negative work must be eliminated, but safe shock absorption must have some component of negative work. In fact, during level running the net negative work done by the muscles must equal the net positive work. Again, how much is a safe minimum is not known. Thus, in the short term, researchers are faced with having to mimic the successful athlete which requires that they establish a detailed and reliable profile of the biomechanical patterns of the winners. The profiles of the near-elite can then be compared with those of the top athlete, and, hopefully, the differences will pinpoint specific changes needed in the movement pattern. In the long term, more detailed and more valid biomechanical models must be developed so that the movement can be synthesized and an ideal be achieved. Thus, specific changes in the motor and kinematic patterns will be suggested to researchers and coaches, who will then be challenged to see if these changes can be implemented in their elite athletes.

References

Asmussen, E.; Bonde-Peterson, F.: Apparent efficiency and storage of elastic energy in human muscles during exercise. Acta physiol. scand. *92:* 537–545 (1974).

Bresler, B.; Berry, F.R.: Energy and power in the leg during normal level walking. Prosthetic Device Research Project, series 11, issue 15. Institute of Engineering Research (University of California, Berkeley, 1951).

Cavagna, G.A.; Margaria, R.: Mechanics of walking. J. appl. Physiol. *21:* 271–278 (1966).

Cavagna, G.A.; Komarek, L.; Mazzoleni, S.: The mechanics of sprint running. J. Physiol. *217:* 709–721 (1971).

Cavagna, G.A.; Kaneko, M.: Mechanical work and efficiency in level walking and running. J. Physiol. *268:* 467–481 (1976).

Chapman, A.E.; Caldwell, G.E.: Factors determining changes in lower limb energy during swing in treadmill running. J. Biomech. *16:* 69–77 (1983).

Crowninshield, R.D.: Use of optimization techniques to predict muscle forces. Trans. ASME J. biomech. Engng. *100:* 88–92 (1978).

Elftman, H.: Forces and energy changes in the leg during walking. Am. J. Physiol. *125:* 339–356 (1939a).

Elftman, H.: The function of muscles in locomotion. Am. J. Physiol. *125:* 357–366 (1939b).

Fenn, W.C.: Frictional and kinetic factors in the work of sprint running. Am. J. Physiol. *92:* 583–611 (1929).

Hardt, D.E.: Determining muscle forces in the leg during normal human walking, an application and evaluation of optimization methods. Trans. ASME J. biomech. Engng. *72:* 72–78 (1978).

Inman, V.T.: Human locomotion. J. Can. med. Ass. *94:* 1047–1054 (1966).

Patriarco, A.; Mann, R.W.; Simon, S.R.; Mansour, J.: An evaluation of the approaches of optimization models in the prediction of muscle forces during human gait. J. Biomech. *14:* 513–525 (1981).

Quanbury, A.O.; Winter, D.A.; Reimer, G.D.: Instantaneous power and power flow in body segments during walking. J. human. Movement Stud. *1:* 59–67 (1975).

Ralston, H.J.; Lukin, L.: Energy levels of human body segments during level walking. Ergonomics *12:* 39–46 (1969).

Robertson, D.G.E.; Winter, D.A.: Mechanical energy generation, absorption and transfer amongst segments during walking. J. Biomech. *13:* 845–854 (1980).

Seireg, A.; Arvikar, R.J.: The prediction of muscular load sharing and joint forces in the lower extremities during walking. J. Biomech. *8:* 89–102 (1975).

Smith, A.J.: The kinetic energy of the human body. J. human. Movement Stud. *1:* 13–18 (1975).

White, S.C.; Winter, D.A.: Mechanical power analysis of the lower limb musculative in race walking. Int. J. Sport Biomech. *1:* 15–24 (1985).

Williams, K.R.; Cavanagh, P.R.: A model for the calculation of mechanical power during distance running. J. Biomech. *16:* 115–128 (1983).

Winter, D.A.; Quanbury, A.O.; Reimer, G.D.: Analysis of instantaneous energy of normal gait. J. Biomech. *9:* 253–257 (1976a).

Winter, D.A.; Quanbury, A.O.; Reimer, G.D.: Instantaneous energy and power flow in gait of normals; in Komi, International series on biomechanics, vol. 4A. Biomechanics V-A, pp. 334–340 (University Park Press, Baltimore 1976b).

Winter, D.A.: A new definition of mechanical work done in human movement. J. appl. Physiol. *46:* 79–83 (1979).

Winter, D.A.: Energetics of human movement, walking and running. Austr. J. Sport Sci. *2:* 26–32 (1982).

Winter, D.A.: Energy generation and absorption at the ankle and knee during fast, natural and slow cadences. Clin. Orthop. rel. Res. *197:* 147–154 (1983a).

Winter, D.A.: Moments of force and mechanical power in slow jogging. J. Biomech. *16:* 91–97 (1983b).

Winter, D.A.: Concerning the scientific basis for the diagnosis of pathological gait and for rehabilitation protocols. Physiother. Canada *37:* 245–252 (1985).

Dr. D.A. Winter, Department of Kinesiology, University of Waterloo,
Waterloo, Ont. N2L 3G1 (Canada)

Med. Sport Sci., vol. 25, pp. 46–57 (Karger, Basel 1987)

Measurement of Human Power Output in High Intensity Exercise

H. K. A. Lakomy

Department of Physical Education and Sport Science, University of Technology Loughborough, Leics., UK

In a dynamic system a mechanical engineer calculates power from the product of force and velocity or from torque and angular velocity, whereas an electrical engineer defines power as the product of potential difference, current and phase angle. Although these definitions appear to be different, they are precise, unambiguous and compatible and result in power being measured in the same unit – the watt. The common factor behind all definitions of power is that they all measure the rate of transfer of energy. All the definitions are interchangeable because the total energy in a system is always conserved.

The measurement of human power in exercise requires the determination of either the rate of chemical energy production within the body or the rate of energy dissipation into the external load. These two energy rates are linked by the mechanical efficiency of the system. Often in sporting situations neither the energy produced nor the energy dissipated is readily measurable.

Some of the difficulties encountered are listed below:

(a) Energy for muscular contraction is produced from a combination of anaerobic and aerobic metabolism. Each source is different in its efficiency and rate of energy production, both of which combine in different proportions depending on the intensity and duration of the activity. In addition, there are several possible energy substrates such as carbohydrate and fat.

(b) The mechanical efficiency of the body is not constant. It is dependent on such variables as stride length in running [Gaesser and Brooks, 1975], stroke rate in rowing/canoeing, pedalling rate in cycling, on enviromental conditions such as ambient temperature, pressure and humidity, as well as on the state of recovery, fatigue or injury and the body composition/somatotype of the individual. Efficiency is also constrained by mechanical restrictions such as the crank length in cycling [Inbar et al., 1983], the weight of the projectile and the shoe-ground interface, and is dependent on whether the activity is continuous or intermittent.

(c) The effects of external forces are often difficult to determine except in those activities such as jumping where the only external force is due to the acceleration of gravity. In many activities the external forces are complex and their influence on the total work done is not readily measured.

(d) The energy required to overcome air or water resistance is dependent both on body speed and posture, which are constantly varying in many activities, as well as on temperature, pressure, wind direction, etc.

(e) There is a time lag between the start of the activity and the detection of the physiological response. Explosive activities such as jumping, throwing, hitting, etc., are often over before the noninvasive techniques for measuring the physiological demands of the activity begin to detect change.

A further complication is that power output in man decreases rapidly with increasing duration of effort. Figure 1 shows the well-established relationship between the duration of an activity and the maximum value of

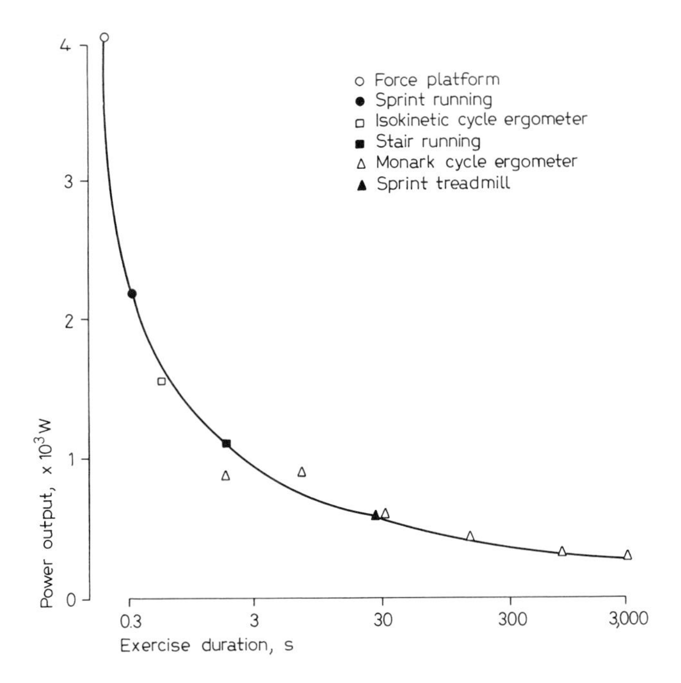

Fig. 1. Relationship between maximum power output and exercise duration for various activities.

power output, for a number of different activities which place a heavy demand on the leg extensors such as jumping, sprinting and cycling. Clearly the power output drops of rapidly as the duration of the performance increases, reaching a fairly constant value for activities lasting longer than 120 [Davies, 1971].

As a consequence of the difficulty in measuring the actual power in sporting activities, exercise physiologists often use so-called pseudopower [Norman et al., 1976] values. Variables which are highly correlated with power such as running speed, stepping frequency and height, speed of swimming and cycling, etc., are measured and used as indicators of the 'work done' or 'power output'.

A series of research techniques have been developed in an attempt to measure the actual power output during various activities. Some of these techniques are outlined below in order of duration of the performance.

The most widely accepted method for measuring the explosive power output of the body is the analysis of the vertical jump from both feet from a force platform, either from a standing start [Davies and Rennie, 1968] or from a drop height (plyometric jump). The instantaneous values of the reaction throughout the take-off phase are logged. The acceleration is determined from the net force (reaction force minus body weight) divided by the mass of the subject. The velocity is obtained from the integration of the area under the acceleration curve. The values for instantaneous power are calculated from the product of force and velocity.

Typical profiles of reaction forces, power and speed are shown in fig. 2. Examination of the timing of the peak values indicates that peak force occurs before peak power, which in turn occurs before the centre of gravity is moving at peak velocity. This relationship is true for all mechanical systems which are free to accelerate.

A test which is of slightly longer duration (2–5 s) is the step-running test devised by Margaria et al. [1966]. From a flying start the subjects run up a flight of stairs as quickly as they can. The power output is calculated from the product of the weight of the subject and the height of the steps (usually the height of the 4th to 6th step) divided by the time taken to cover the steps investigated. This method of calculation assumes that the horizontal component of the running speed is constant over the steps investigated and that the rise in the centre of gravity is the same as the height of the steps.

Several different methods have been developed for the calculation of mechanical power output during running. These include representing the body as only a single point mass [Fukunaga et al., 1978], as a single point

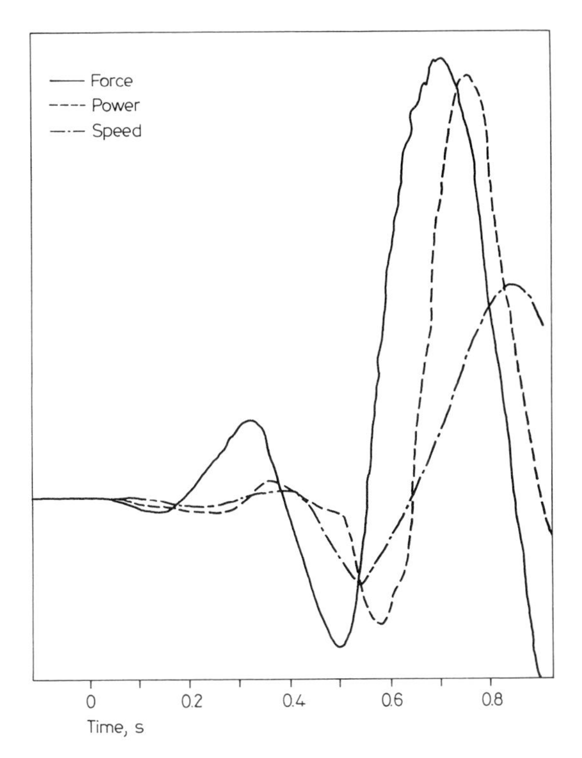

Fig. 2. A typical relationship between force, power and speed during a standing jump on a force platform.

mass in conjunction with the movements of the limbs relative to that point [Fenn, 1930; Cavagna et al., 1964], or as a series of rigid links or segments. Several segmental models which examine the segmental energy levels, joint power and energy transfer, namely the 'internal' work, have been put forward. These include the analysis procedure termed pseudowork proposed by Norman et al. [1976]. Using this technique the absolute changes in the instantaneous energy of a segment's potential, translational kinetic and rotational kinetic energies are summed together. This technique has been criticized as not allowing for energy exchange between energy forms within a segment [Pierrynowski et al., 1980]. Other segmental models attempt to account for energy transfers both within and between segments [Winter, 1978, 1979].

All the techniques require 3-D cinematography for instantaneous segment, or centre of mass, velocity determination and are usually used in conjunction with force analysis using force platforms. Many assumptions have to be made in the calculations such as the amount of the total negative work that can be attributed to muscular work, the effects of the storage and recovery of elastic energy, and the relative metabolic cost of positive and negative work. Estimations must also be made of limb length and mass.

Williams and Cavagna [1983] examined several of the computational techniques and obtained a very wide range of values for mechanical power depending upon the particular assumptions made and the technique employed. As yet there is no universal agreement on which computational method is correct.

The most widely used protocols for both the biomechanical and the physiological assessment of endurance are based on the cycle ergometer. More recently, tests for evaluating short-term explosive power have also been devised using the cycle ergometer [Ayalon et al., 1975; Bar-Or, 1978]. The external loading on such ergometers are from electromechanical devices, electromagnetic or eddy-current brakes, air resistance or the most commonly adopted method of friction loading. The friction-loaded ergometers are based on the basic design by von Dobeln [1954]. Some of the ergometers [Sargeant and Davies, 1977] limit the pedalling speed to a constant predetermined value (isokinetic) whilst the majority allow acceleration. All the ergometers measure the external power generated by the subject.

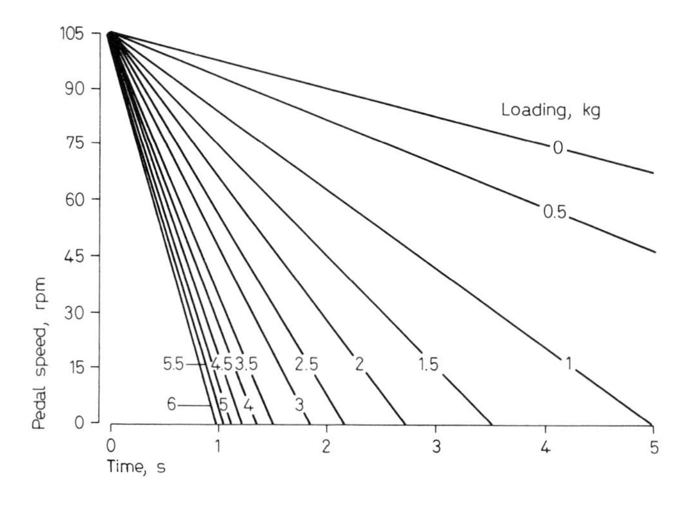

Fig. 3. Plot of time taken for the flywheel to decelerate from 105 to 0 pedal rpm under different frictional loads.

On the friction-loaded ergometers the power is conventionally calculated from the product of the resistive load and the flywheel speed. The calculation of the work done and the power output in this way assumes that the flywheel is either revolving at a constant angular velocity or has no moment of inertia. In these tests, however, neither of these conditions are satisfied. In order to measure the true power output in such sprint activities, both the work done in accelerating the flywheel and the energy used to overcome the friction generated by the resistive load must be calculated and summed together [Lakomy, 1986]. In order to measure the work being done in accelerating the flywheel, the 'acceleration balancing load', defined as the load which would be required to prevent the acceleration, must be determined. To do this, the speed of the flywheel must be constantly monitored, preferably by a computer. This can be done by the flywheel driving a small DC generator whose output is connected to the computer via an analogue-to-digital converter. Fig. 3 shows a set of deceleration curves resulting from applying known resistive loads to the previously freely rotating unloaded flywheel. If the deceleration is then plotted against the applied load (fig. 4) then a regression equation of the form:

$$Y = mX + C$$

is obtained where Y is the acceleration, X is the load, and m and C are constants. If X load produces Y deceleration then to prevent acceleration of

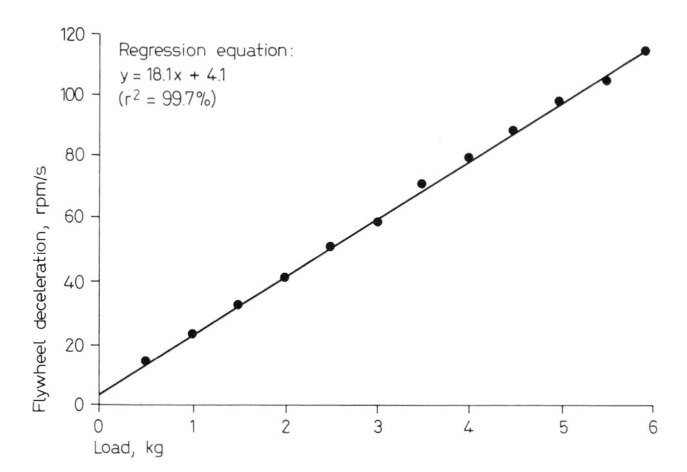

Fig. 4. Plot of flywheel deceleration against load.

the flywheel at a rate Y, load X must be added to the resistive load. If the flywheel is decelerating, e.g. when the subject is fatiguing, then X must be subtracted from the resistive load. As the value of Y is constantly changing (but can be calculated by the computer) then the value of X must also be varying. It is this load value that is the 'acceleration balancing load'. The instantaneous power output is then calculated from:

power output = flywheel speed × (resistive load + acceleration balancing load).

Tests have clearly shown [Lakomy, 1986] that when no correction is made for the work done in accelerating the flywheel then (1) peak power output is greatly underestimated (up to 65%); (2) the time taken to reach peak power output is overestimated, and (3) the instantaneous values of power output at any time during the test are incorrect.

Several studies have examined the load that should be applied to achieve the highest value for peak power output [Evans and Quinney, 1981; Nakamura et al., 1985]. It is not surprising that in such studies the value found for this optimum load was very high, around 100 g/kg body weight, which approaches the maximum loading that the subjects could tolerate without 'stalling'. The greater the applied load, the slower will be the acceleration of the flywheel when the subject applies the propulsive torque. The component not corrected for using the conventional method of calculation is the magnitude of the work done in accelerating the flywheel. As this is directly proportional to the acceleration then the smaller the acceleration the smaller will be the error. If, however, the correct method of calculation is used then the value of peak power output measured appears to be independent (within limits) of the load applied [Lakomy, 1984].

By using the correct method of calculation the instantaneous value of applied torque can be determined if the instantaneous value of acceleration of the flywheel is known. This is of particular interest to those researchers attempting to instrument the cranks using slip rings for intra-pedal torque profiles as this technique allows for a simple indirect method of obtaining such data.

Sprint running in contrast to sprint cycling is a weight-bearing activity. The cycle ergometer may, therefore, be of limited value to those interested in the evaluation of sprint running. Recently, an ergometer for the assessment of power generation has been developed [Lakomy, 1985], based on a nonmotorized treadmill. Fukunaga et al. [1978] and Cavagna et al. [1965] have shown that the work done in raising the centre of gravity during the stride cycle is independent of running speed whilst the work done in propelling the

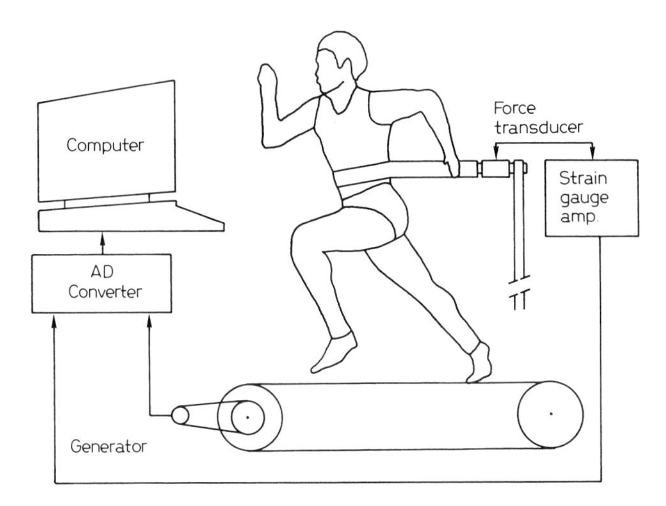

Fig. 5. Diagram of the data collection system for the sprint treadmill.

runner is a function of running speed. The sprint ergometer utilizes this result and measures only the propulsive component, i.e. the horizontal component of the applied force and power output.

The sprint treadmill ergometer allows the subjects to run at speeds similar to those achieved in free running. It also permits the same variability in instantaneous work rate during sprinting that the cycle ergometer allows during cycling. The instantaneous values of power output can be determined throughout the sprint.

The commercially available nonmotorized treadmill (Woodway Model AB) was modified in the following way so that measurements could be made. The treadmill, which normally slopes backwards, was levelled and anchored. A high precision generator was mounted so as to measure the speed of the treadmill belt. The tether belt which was passed around the subject's waist was attached to a force transducer (Pioden Ltd., Model UF2) which was attached to an adjustable, but rigid, crossbar. The output from the strain gauge amplifier and the speed generator was continuously monitored by a BBC model B microcomputer via a multiplexed multichannel 12-bit analogue-to-digital converter. Figure 5 shows a schematic diagram of the data collection system.

The product of treadmill belt speed and the applied force is defined as the propulsive power. This calculation assumes: (a) that the error resulting from the points of force application and measurement not being the same, is small; (b) that the movement of the tether belt (variation from the horizontal) does not introduce a large error; (c) that the error due to elasticity in the tethering system is small, and (d) that little of the subject's weight is detected as a horizontal force due to forward lean.

After each test the following information was calculated and displayed: (a) the mean propulsive power for each second of the sprint; (b) the mean propulsive power for the test period; (c) the total propulsive work done, and (d) a fatigue index defined as the difference between the peak and the lowest (or end) propulsive power output values defined as a percentage of the peak propulsive power.

Results from the treadmill show that during a sprint from either a stationary or jogging start peak force occurs before peak power and in turn peak power precedes peak speed (fig. 6). This result agrees with the mechanical principles described earlier.

Further tests have shown that the propulsive power required to maintain constant submaximal running speeds, for each individual, increased linearly with the running speed (r^2 = 93.8%). In addition, both the propulsive force and the propulsive power needed to run at a constant speed increased with the weight of the subject (r^2 = 93.0 and 94.3%, respectively).

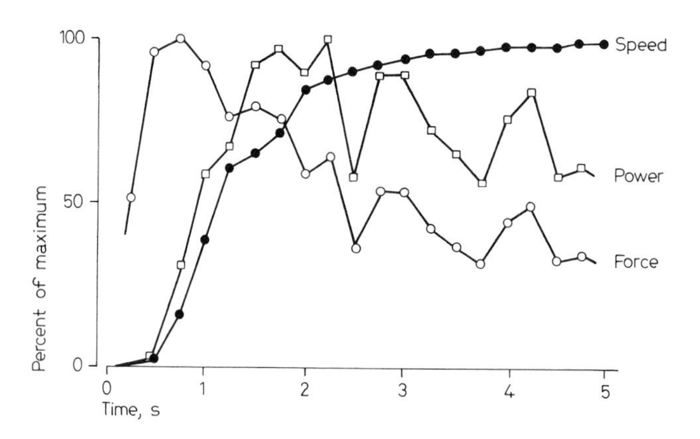

Fig. 6. A typical relationship between force, power and speed during the first 5 of a sprint.

Elasticity in the tethering straps was found to act as a low pass filter smoothing out the force peaks. It is, therefore, important that the system should be made as rigid as possible, without causing discomfort to the runner.

Using video cameras the deviation of the tether straps from the horizontal during a sprint was investigated. Initially the straps were set horizontal when the subject was standing still and upright. It was found that the arc subtended by the straps was approximately 8°, and that the centre of this arc was approximately 7.8° below the horizontal. Prior to each test the tether straps should, therefore, be set to approximately 8° above the horizontal when the subject is standing still, so as to minimize the error introduced by the movement of the straps.

Comparing individuals running on the nonmotorized and motorized treadmills, at a given speed, it was found that the forward lean was more pronounced on the nonmotorized treadmill (approximately 13° at 2.87 $m \cdot s^{-1}$).

Peak propulsive power outputs (mean of 1 second in excess of 1 kW were achieved during the acceleration phase of the sprint. Instantaneous values of peak power output during the stride was found to reach in excess of 3 kW. Fukunaga et al. [1978] have shown that this propulsive power is approximately 80% of the total power output, with the other 20% required for the work against gravity. These results show that the power output measured on the sprint treadmill are similarly related to duration of activity as were those generated by other forms of exercise.

Although not identical with free sprinting, sprint running on the nonmotorized treadmill is very similar. In this paper only the propulsive forces and power outputs produced on the treadmill have been examined; however, it is possible to measure the instantaneous values of running speeds, stride frequency, stride length, contact time, and flight time, without any modification to the system, if a fuller biomechanical analysis is required.

In conclusion, is has been shown that when the ergometer does not prevent acceleration, peak force generation precedes peak power output, and in turn peak power output occurs before peak speed. In addition, during exercise the level of power output that can be sustained by the body, is a function of the duration of the activity, the longer the duration the lower the power output.

Power output is often difficult to measure in exercise; however, if both the value of the force being generated by the body and the velocity of its point of application can be measured, then the instantaneous product can be calculated. Although the ergometers described in this paper all measure power output using different techniques, they all produce values in the same unit – the watt.

References

Ayalon, A.; Inbar, O.; Bar-Or, O.: Relationships among measurements of explosive strength and anaerobic power. Biomechanics *IV:* 572–577 (1975).

Bar-Or, O.: A new anaerobic capacity test – characteristics and applications. Proc. 21st Wld Congr. Sports Med., Brasilia 1978.

Cavagna, G. A.; Margaria, R.; Arcelli, R.: A high speed motion picture analysis of the work performed in sprint running. Res. Fim. *5:* 309–319 (1965).

Cavagna, G. A.; Saibene, F. P.; Margaria, R.: Mechanical work in running. J. appl. Pphysiol. *19:* 249–256 (1964).

Davies, C. T. M.: Human power output in exercise of short duration in relation to body size and composition. Ergonomics *14:* 245–256 (1971).

Davies, C. T. M.; Rennie, R.: Human power output. Nature, Lond. *217:* 770–771 (1968).

Dobeln, W. von: A simple bicycle ergometer. J. appl. Physiol. *7:* 222–224 (1954).

Evans, J.; Quinney, H.: Determination of resistance settings for anaerobic power testing. Can. J. appl. Sports Sci. *6:* 43–56 (1981).

Fenn, W. O.: Frictional and kinetic factors in the work of sprint running. Am. J. Physiol. *92:* 583–611 (1930).

Fukunaga, T.; Matsuo, A.; Yuasa, K.; Fujimatsu, H.; Asahina, K.: Mcehanical power output in running. Biomechanics *VI-B:* 17–22 (1978).

Gaesser, G. A.; Brooks, G. A.: Muscular work efficiency during steady-rate exercise: effects of speed and work rate. J. appl. Physiol. *38:* 1132–1146 (1975).

Inbar, O.; Dotan, R.; Trousil, T.; Dvir, Z.: The effect of bicycle crank-length variation upon power performance. Ergonomics *26:* 1139–1146 (1983).

Lakomy, H. K.: An ergometer for measuring the power generated during sprinting. J. Physiol. *354:* 33P (1984).

Lakomy, H. K.: Effect of load on corrected peak power output generated on friction-loaded cycle ergometers. J. Sports Sci. *3:* 240 (1985).

Lakomy, H. K.: Measurement of work and power output using friction-loaded cycle ergometers. Ergonomics *29:* 509–514 (1986).

Margaria, R.; Aghemo, P.; Rovelli, E.: Measurement of muscular power output (anaerobic) in man. J. appl. Physiol. *21:* 1662–1664 (1966).

Nakamura, Y.; Yoshiteru, M.; Miyashita, M.: Determination of the peak power output during maximal brief pedalling bouts. J. Sports Sci. *3:* 181–187 (1985).

Norman, R.; Sharratt, M.; Pezzack, J.; Noble, E.: Re-examination of the mechanical efficiency of horizontal treadmill running. Biomechanics *V-B:* 87–93 (1976).

Pierrynowski, M. R.; Winter, D. A.; Norman, R. W.: Transfer of mechanical energy within the total body and mechanical efficiency during treadmill walkding. Ergonomics *212:* 39–46 (1980).

Sargeant, A. J.; Davies C. T. M.: Forces applied to cranks of a bicycle ergometer during one- and two-leg cycling. J. appl. Physiol. *42:* 514–518 (1977).
Williams, K. R.; Cavagna, P. R.: A model for the calculation of mechanical power during distance running. Biomechanics *16:* 115–128 (1983).
Winter, D. A.: Calculation and interpretation of the mechanical energy of movement. Exer. Sports Sci. Rev. *6:* 183–201 (1978).
Winter, D. A.: A new definition of mechanical work done in human movement. J. appl. Physiol. *46:* 79–83 (1979).

H. K. A. Lakomy, MD, Department of Physical Education and Sport Science, University of Technology, Loughborough, Leics. LE11 3TE (UK)

Med. Sport Sci., vol. 25, pp. 58–71 (Karger, Basel 1987)

Biomechanical Limitations to Sprint Running

Graeme A. Wood[1]

Department of Human Movement Studies,
University of Western Australia, Nedlands, Australia

"Now bid me run,
And I will strive with things impossible.
Yea, get the better of them."
W. Shakespeare (*Julius Caesar*, Act II, Scene 1)

Introduction

The objective in sprint running is clearly to traverse the ground as quickly as possible, and this is one activity in which man is not as adept as are other members of the animal kingdom. Further, it is an activity that is not without hazard in that the performer occasionally exceeds the mechanical limits of his musculo-skeletal system and a muscle is torn or other structures are damaged. In the realm of human athletic endeavor sprint running has long been one of the most popular events; an eminence that is demonstrated by early Greek art and the voluminous writings on how to run faster. Despite all this attention, the essential factors that limit one's sprinting ability are still not clearly understood, nor are the mechanisms for performance enhancement or injury avoidance. This chapter presents some recent research that has focussed on these questions, the major emphasis being on the mechanics of the lower extremity once maximal speed has been attained. While good starting and finishing technique and efficient arm actions are undoubtedly important, the latter more for balancing purposes, these factors are not considered here. A more general discussion of the biomechanics of running can be found in two other reviews that have recently been published [Vaughan, 1984; Williams, 1985], but little in these pertains to sprint running.

[1] The author gratefully acknowledges the contribution that the following people have made to the personal research reported in this chapter: Phillip De La Hunty, Andrew Cresswell, Geoffrey Strauss, John Anglim, and Dr.'s Robert Marshall and Les Jennings. This research has been supported in part by grants from the Sir Robert Menzies Foundation for Health, Fitness and Physical Achievement.

Basic Factors in Sprint Running

There are many biomechanical factors that govern man's sprinting ability. Ultimately sprinting speed is simply the product of stride length (the distance travelled between successive foot strikes on the same side of the body) and stride frequency, but these variables are merely the outcome of the successful integration of many mechanical and neuromuscular processes, as is depicted in figure 1.

At speeds approaching maximum it is stride frequency that undergoes most change (fig. 2). Not only does this variable increase but the athlete spends proportionately less time in contact with the ground [Dillman, 1975; Luhtanen and Komi, 1978]. A likely consequence of this would be that less impulse could be generated by the propulsive ground reaction forces and the resulting gain in momentum would be sufficient only to combat the retard-

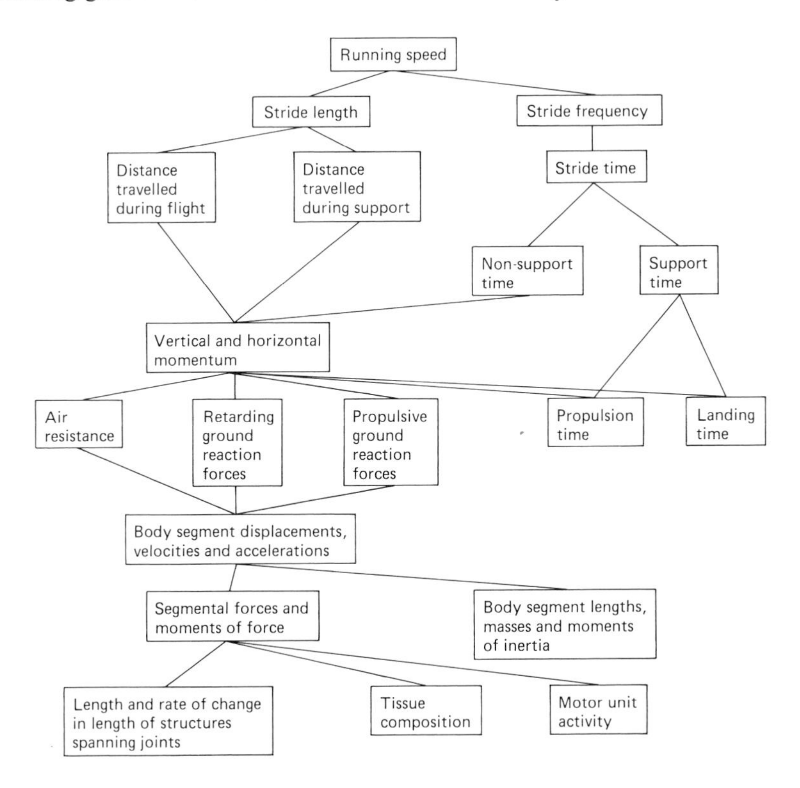

Fig. 1. Schema of biomechanical factors in sprint running. Adapted from Hay and Reid [1982].

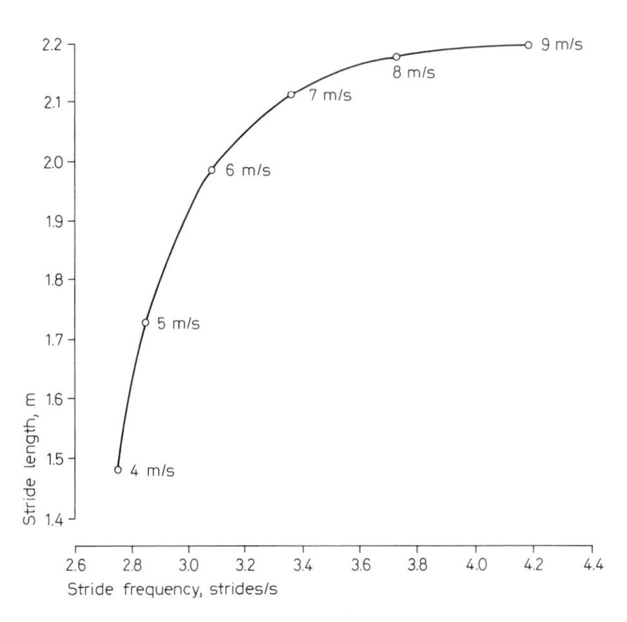

Fig. 2. Relationship between stride length, stride frequency and speed of running. Adapted from Dillman [1975].

ing influences of air resistance and foot strike (fig. 3). It can be argued, however, that, since support time is being reduced when approaching maximal velocity, the rate-limiting aspect is the leg recovery, an aspect which perhaps cannot be reduced without reducing stride length or producing unwanted retarding impulses at foot strike. Certainly world-class sprinters, when compared to decathletes, have been found to have higher stride frequencies, and these are achieved through briefer ground contact times [Kunz and Kaufmann, 1981]. Also, recent comparisons between Olympic 200-meter finalists indicate that the more successful sprinters have higher stride frequencies, and again it is a briefer support time that is the determining factor [Mann and Herman, 1985].

It appears that the better sprinter not only minimizes retarding forces at foot strike but also generates propulsive forces earlier during ground contact in order to be airborne again with a minimum of delay. This suggests that the critical phase in sprint running is during late leg recovery and early support. The optimal leg recovery pattern is known to be characterized by good knee flexion and a high knee lift [Sinning and Forsyth, 1970], a rapid foot descent with the foot moving rapidly backwards (relative to the body) as it strikes the

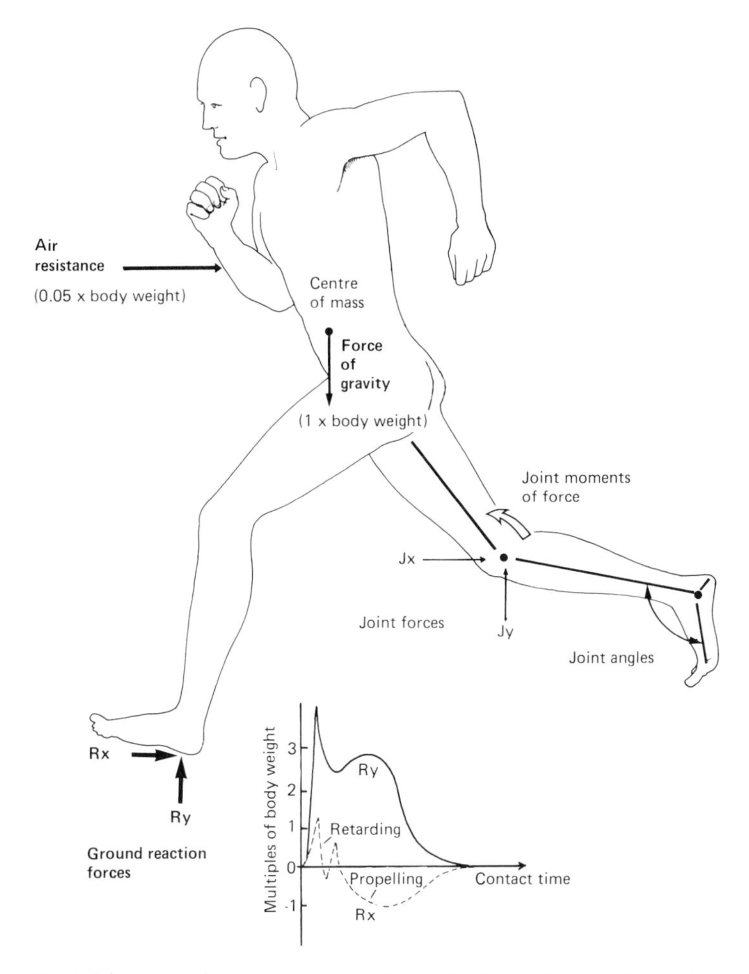

Fig. 3. Diagrammatic representation of the mechanical variables which ultimately determine sprinting speed, showing the relative magnitude of external forces acting on the body, and the segmental variables which produce these effects.

ground [Mann and Herman, 1985], and contact made at a point almost beneath the body's center of gravity [Deshon and Nelson, 1963; Kunz and Kaufmann, 1981; Mann and Herman, 1985]. In essence, the good sprinter propels himself forward early in the support phase as the support limb is pulled under him rather than pushing from a rearward leg position. In fact, the better sprinters do not even extend their knee fully before take-off [Mann and Herman, 1985].

Kinetic Analyses of the Lower Extremity Action in Sprint Running

What are the muscular requirements of these lower limb actions? Several investigators have recently quantified the forces produced in the lower extremity during sprint running, and in the author's own laboratory independent indices of muscular activity have also been obtained. The methods used are depicted in figure 4 and require the simultaneous capture of high speed film and force platform data, together with electromyographic (EMG) activity from active muscles. An analytical model can then be established which enables resultant joint moments of force to be determined using the inverse dynamics approach given (i) segmental displacements, velocities and accelerations obtained from the film record, and (ii) anthropometric data available from cadavers [Wood, 1977]. Briefly, the approach utilises two equations that define the translational dynamics of a body segment, namely

$$\Sigma F_j^x = m_j \cdot a_j^x ,$$

and

$$\Sigma F_j^y = m_j \cdot g + m_j \cdot a_j^y ,$$

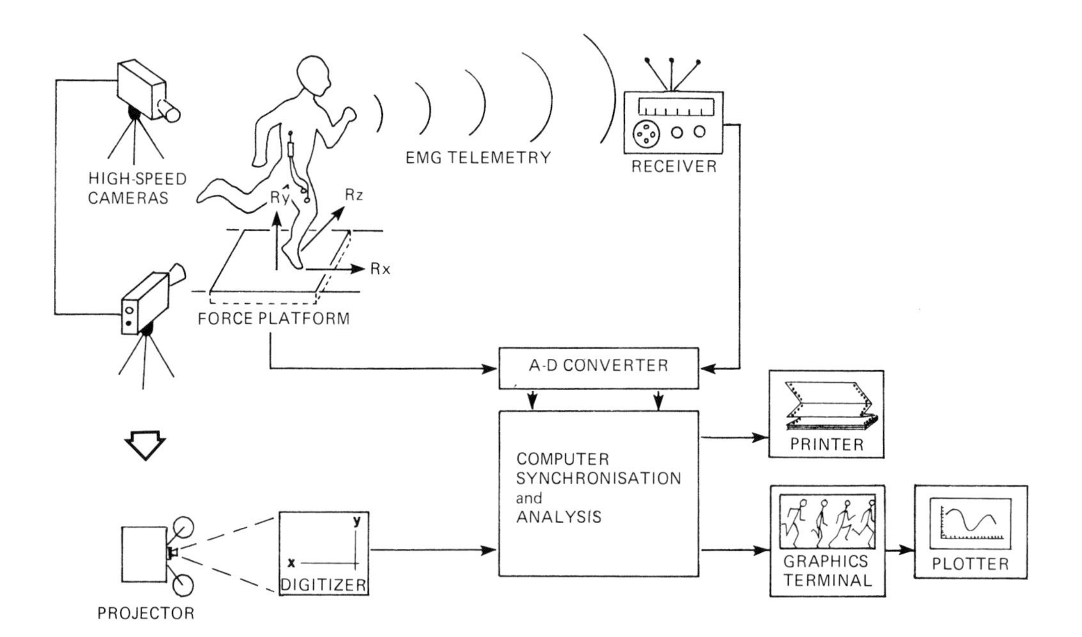

Fig. 4. Methodological approach used in biomechanical analyses of sprint running.

and one equation defining the rotational dynamics of that body segment, viz:

$$\Sigma M_j = I_j^{Gz} \cdot \alpha_j ,$$

where F_j^x and F_j^y are the x and y components of the reaction forces impinging on the proximal and distal ends of the j-th segment, and the M_j's are the moments of force due to these reactions as well as the turning effects of the net muscle pull across the proximal and distal joints. The solution to these equations relies on a knowledge of the segment's mass (m_j), location of mass center, mass moment of inertia (I_j^{Gz}), as well as measures of the linear (a_j) and angular (α_j) acceleration of the body segment in the x-y (saggital) plane (g is the constant acceleration due to gravity). Typical results from such an analysis are shown in figure 5 for a group of 9 male athletes. Electromyographic activity has been full-wave rectified and, together with kinetic data, time-averaged and normalised to provide an ensemble average over all subjects. Muscle power at each joint has been calculated as the product of joint moment of force and joint angular velocity, i.e.:

$$P = M_p \cdot (\omega_d - \omega_p),$$

where M_p is the resultant moment of force due to muscle action across the proximal joint, and ω_p and ω_d are the absolute angular velocities of the proximal and distal segments, respectively [Winter et al., 1976]. For clarity, an extensor joint moment of force has been designated as positive in figure 5, as is all energy-generating muscle power.

Good agreement can be seen between joint moment of force patterns and EMG activity (when adjustments are made for electromechanical delays, i.e. 20–40 ms), despite the fact that the former are only indices of muscle predominance and seldom indicate the absolute magnitude of muscular effort. The dominant positive moment of force at the ankle joint during ground contact indicates the obviously important role of the ankle plantar flexors: firstly acting eccentrically (negative power) to absorb energy and then concentrically thereby generating (or releasing) energy later in the support phase. This active stretch-shortening cycle, coupled with the early cessation of m. triceps surae EMG activity well before the diminution of ankle plantar flexor moments suggests that elastic-strain energy is being stored in this musculo-tendinous unit during the early part of support and recovered toward toe-off. M. triceps surae are particularly well adapted for this purpose due to their long tendon and the ability of this tissue to return as much

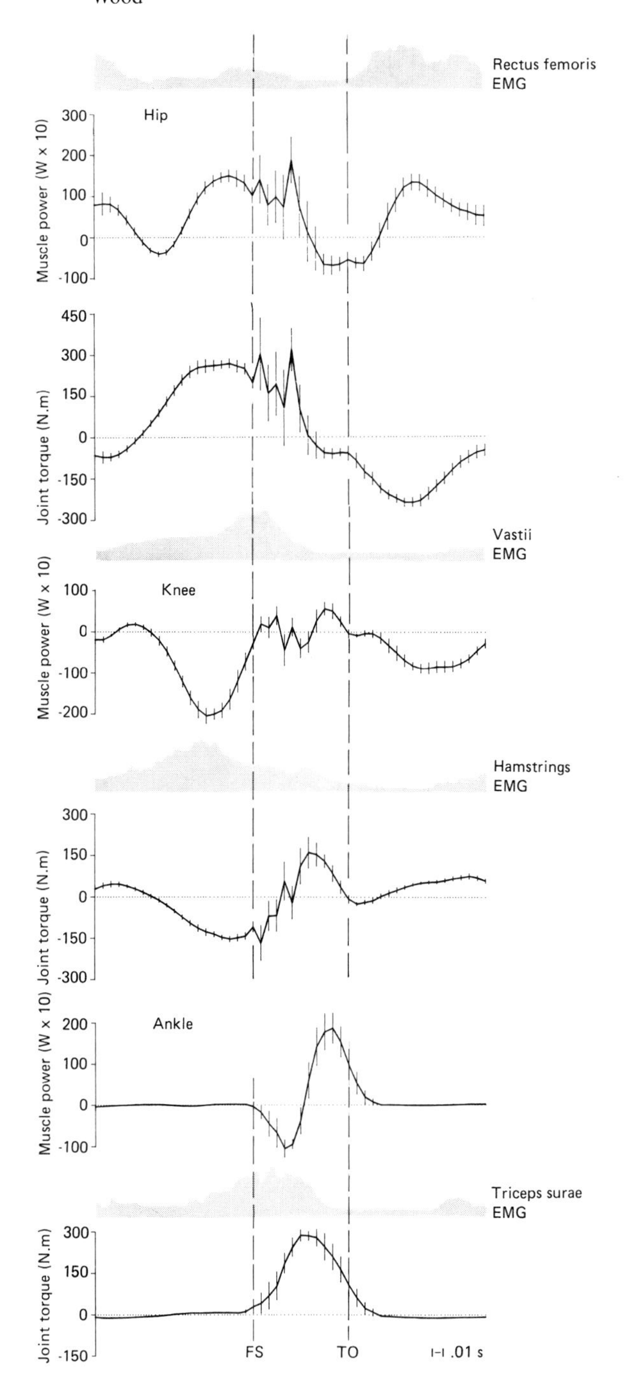
Rectus femoris
EMG
Hip
Muscle power (W x 10)
300
200
100
0
-100
Joint torque (N.m)
450
300
150
0
-150
-300
Vastii
EMG
Knee
Muscle power (W x 10)
100
0
-100
-200
Hamstrings
EMG
Joint torque (N.m)
300
150
0
-150
-300
Ankle
Muscle power (W x 10)
200
100
0
-100
Triceps surae
EMG
Joint torque (N.m)
300
150
0
-150
FS
TO
.01 s

as 93% of the energy stored in it (approximately 33 J) [Alexander, 1984, 1986]. Efficiencies achieved in these ways purportedly account for a substantial fraction of the work done during each step (at push-off) in sprint running [Cavagna, 1974].

Muscle power analysis also reveals the predominance of eccentric muscle activity in the knee flexors before foot strike and, to a lesser extent, in the extensors during mid-support. During forward swing the knee flexors contract powerfully to control the momentum of the leg and to prepare for an efficient foot strike. In so doing they absorb a considerable amount of energy (approximately 125 J). This energy absorption accounts for the deficit observed between energy input at the hip and total limb energy [Chapman and Caldwell, 1982a]. The period of knee flexor muscle activity extends briefly into the stance phase with energy now being generated and transferred to the shank. The dominance of knee flexor activity early in the support phase might seem surprising but has been interpreted by Mann and Sprague [1980] as an attempt by the sprinter to minimise the braking action at touchdown by pulling the body forward and over the touchdown point during initial ground contact. However, during this phase widely fluctuating anterior-posterior ground reaction forces are often recorded and if entered into segmental calculations will produce highly variable moments of force at knee and hip. Whether or not these effects are actually accommodated by the structures spanning a joint is not known, and they may equally be just manifestations of foot slippage. It may therefore be inappropriate to include these high-frequency force effects in kinetic calculations (as was done here) when kinematic measures are low-pass filtered (to minimise measurement error). Nevertheless, both the hamstrings and m. triceps surae muscle groups display EMG activity well into the support phase and both could be effecting a knee flexor moment, although there is clearly some degree of co-contraction at the knee.

By mid-support the knee extensors dominate, firstly with a brief period of energy absorption followed by concentric activity which generates, as well as effects, a transfer of energy from the shank to the thigh. Early leg recovery is characterised by energy absorption by the knee extensors, clearly achieved by m. rectus femoris insofar as the vastii are only active during the support phase (assuming a 20 to 40-ms electromechanical delay). Similarly, the thigh's momentum after toe-off is arrested by eccentric activity of the hip flexors, which then initiate the forward swing.

Fig. 5. Average kinetic and muscular (EMG) actions at ankle, knee and hip joints during one complete stride in sprint running. FS = foot strike; TO = toe-off. For convenience, extensor moments of force are designated as positive, as is all energy-generating muscle power.

The utility of two-joint muscles is clearly seen in the pattern of hip moments of force. The forward rotation of the thigh is facilitated by both the generation and transfer of energy from the hip flexors at a time when the energy of the shank must also be absorbed. Similarly, the later concentric activity of the hip extensors retards the thigh while the knee flexors absorb energy from an otherwise flailing shank. This dual role of the hamstring muscle group is, however, not without its potential hazards. The magnitude of the hip and knee power requirements indicates that the hamstrings are under some considerable stress, a factor exacerbated by attempts to minimise retarding forces at foot strike by prolonging the hip extension/knee flexion phase. The magnitude of these moments at touch-down have been found to be positively correlated with the previous occurrence of hamstring injury in sprinters, but are also positively correlated with horizontal velocity loss [Mann and Sprague, 1980; Mann, 1981]. The subsequent period of potential co-contraction (antagonism) between knee and hip extensors also has implications in terms of sprinting efficiency in that propulsion (unlike stability) can best be achieved by the exclusive use of one muscle group [Elftman, 1940].

There is little kinetic evidence to suggest that muscular efforts in the late support phase are essential to sprinting proficiency, nor that this is a high risk period for muscle injury. However, one of the short-comings of joint moment of force and muscle power analyses is that the actions of a two-joint muscle cannot be fully appreciated. For example, simultaneous concentric hip extensor and eccentric knee flexor moments of force, whilst strongly implicating the hamstring muscle group (a supposition here supported by EMG activity patterns), does not provide definitive answers on the mechanical state of this muscle group, i.e. whether it is shortening, being stretched, or acting isometrically (or even isokinetically!). The solution to this problem has particular relevance not only for muscle conditioning for sprinting (e.g. the principle of specificity of exercise), but would also permit the identification of muscular states of mechanical insufficiency or neural inhibition as potentially limiting factors. To achieve this goal, the simple link-system model from which joint moment of force measures were obtained must be endowed with anatomical detail such as that proposed by Frigo and Pedotti [1978]. Here points of origin and insertion and lines of action relative to instantaneous joint centers of rotation have been specified and, when applied to the actions of the lower extremity during sprint running, yield the hamstring muscle length information displayed in figure 6.

It is clear that the hamstring muscle group is undergoing eccentric contraction when it becomes active during late recovery. M. semitendinosus

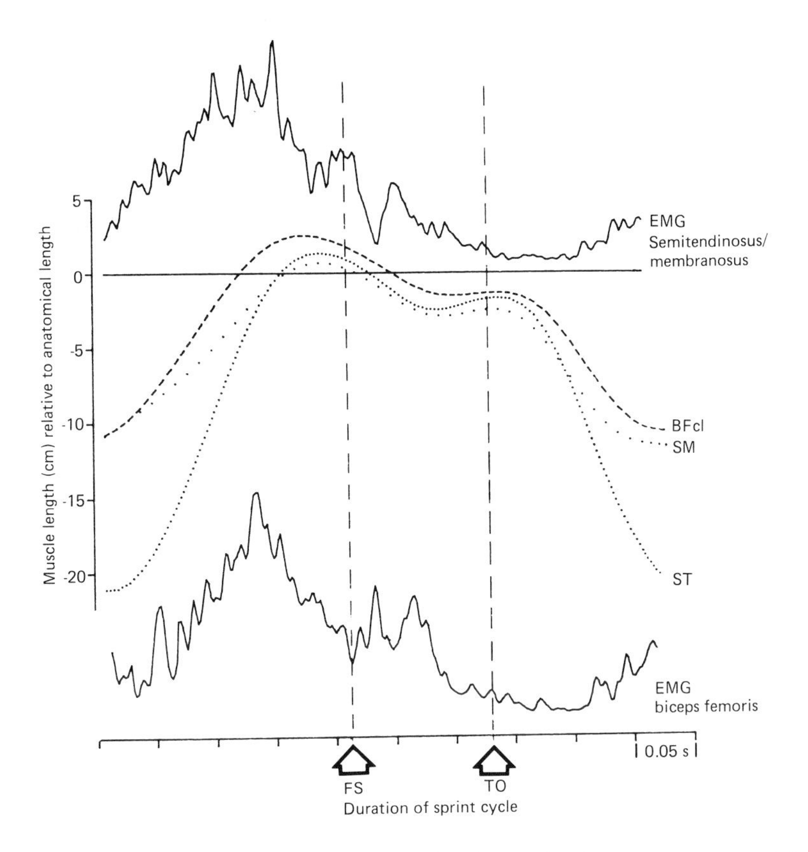

Fig. 6. Hamstring muscle length changes and EMG activity during an average sprinting stride. These data are based on an analytical model proposed by Frigo and Pedotti [1978]. Anatomical lengths correspond to vertical standing lengths. FS = foot strike; TO = toe-off; BFcl = m. biceps femoris capus longus; ST = m. semitendinosus; SM = m. semimembranosus.

sustains the greatest rate of lengthening, while the long head of m. biceps femoris is stretched most. Clinical studies indicate that m. biceps femoris is the most prone to injury [Gray, 1975] and this is not surprising when one considers that this muscle has the shortest fibers per muscle length and smallest pinnation angle of all the components of the hamstring muscle group [Wickiewicz et al., 1983].

The rates of muscle stretching calculated by this model are extremely high and although force output increases with increasing eccentric velocity, the sprinter is operating in the upper reaches of the force velocity curve where the relationship between force and velocity levels off [Komi, 1979] and little additional force production is possible. It is also interesting to note that these muscles reach peak activity at a time (approx. 0.07 s prior to foot-strike) when their respective lengths are just greater than resting length for it is here that the greatest force is physiologically possible.

Performance Enhancement and Fatigue Effects in Sprint Running

It would seem reasonable to suggest that a worthwhile approach to the study of limiting factors in sprint running would be to study sprinters before and again after beneficial training, or during the onset of fatigue under race conditions. In this way, limiting aspects of a sprinter's action would be highlighted, with the possibility that causal factors might be identified.

Studies of the effects of fatigue on sprint running indicate that, as a result of fatigue, stride rate decreases [Bates and Osternig, 1977; Sprague and Mann, 1983], sometimes with a concomitant increase in stride length. The decreases in stride rate appear to be attributable to an increase in both support and non-support times. Decreases in the extremes of flexion and extension have also been reported [Bates and Osternig, 1977; Chapman, 1982], together with increases in the velocity of the foot relative to the ground at foot-strike [Sprague and Mann, 1983]. These findings are consistent with the critical factors discussed above, but it must be noted that considerable individual variability is often observed and it is difficult to know whether these changes are merely manifestations of a slower sprinting speed, or are indeed causal.

To move faster overground is an achievement for which many sports-people strive and many training strategies have been devised in the pursuit of this goal. Traditionally the principles of over-load were applied (e.g. running with weights, running uphill, repetition and interval work), but in 1971 Dintiman advocated the over-speed method whereby the athlete is assisted to speeds greater than his normal maximum. This type of training can take several forms, including: downhill running, downwind running, towing and high-speed treadmill running. The results are usually beneficial, if somewhat transitory [Wood et al., 1982]. Improvements in both stride length and stride rate have been reported [Nelson and Osterhoudt, 1971; Sandwich, 1967; Tinning and Davis, 1978; Mero and Komi, 1985]. In one particular high-speed treadmill study with which the author was associated [Wood et al., 1982], and which resulted in a 0.45 m/s increase in sprinting speed, it was clearly evident that the hamstring muscle group was under considerable stress. An increased contractility of this muscle group was suggested by smaller knee flexion angles during both leg-recovery and early support, and greater peak moments of force for both hip extensors and knee flexors during subsequent overground sprinting. It was also evident that this form of over-speed training must be used cautiously insofar as some subjects had a recurrence of hamstring muscle strain.

In recent years some biomechanical studies of performance enhancement have utilised the mathematical procedures of simulation and optimisation in order to identify a better technique. In one such study [Wood et al., 1986], a computer simulation of a quicker leg recovery was undertaken using an optimal control approach. Of the optimisation (objective) functions examined a minimisation of the sum of joint muscle powers produced the most realistic limb movements, and a reduction in leg recovery time required less knee lift but greater eccentric force moments, particularly about the knee during the terminal period just prior to foot strike. These findings are consistent with the energy analyses undertaken by Chapman and Caldwell [1982b] who conclude that the single most important aspect limiting leg recovery is an inability to increase the peak eccentric muscle moment at the knee prior to foot-strike. This inability to attain large eccentric forces limits the required rate of energy reduction to prepare for foot-strike and thus prolongs recovery time. These findings reinforce the wisdom of eccentric conditioning of the hamstring muscle group as a means of performance enhancement and injury avoidance.

Conclusions

This chapter has reviewed in some detail the lower extremity mechanics of a sprint runner. It has been suggested that the ultimate speed-limiting factor is leg recovery, and that this in turn is largely governed by the contractile capacity of the hamstring muscle group. The mechanical actions and functional state of this muscle group during a sprint cycle have been examined, and it has been suggested that injury-free performance enhancement could be achieved by eccentric conditioning of this muscle group.

References

Alexander, R.M.: Walking and running. Am. Scient. *72:* 348–354 (1984).
Alexander, R.M.: Elastic mechanisms in the movement of animals. Keynote Paper North American Congr. Biomechanics, Montreal 1986.
Bates, B.T.: Osternig, L.R.: Fatigue effects in running. J. Motor Behav. *9:* 203–207 (1977).
Cavagna, G.A.: Elasticity in sprint running, XXth Wld Congr. Sports Medicine Congr. Proc., Melbourne 1974, pp. 107–108.
Chapman, A.E.: Hierarchy of changes induced by fatigue in sprinting. Can. J. appl. Sport Sci. *7:* 116–122 (1982).
Chapman, A.E.; Caldwell, G.E.: Factors determining changes in lower limb energy during swing in treadmill running. J. Biomech. *16:* 69–77 (1982a).
Chapman, A.E.; Caldwell, G.E.: Kinetic limitations of maximal sprinting speed. J. Biomech. *16:* 79–83 (1982b).
Deshon, D.E.; Nelson, R.C.: A cinematographical analysis of sprint running. Res. Q. *35:* 451–455 (1963).
Dillman, C.J.: Kinematic analyses of running; in Wilmore, Keogh, Exercise and sports sciences reviews, vol. 3, pp. 192–218 (Academic Press, New York 1975).
Dintiman, G.B.: Sprinting speed. Its improvement for major sports competition (Thomas, Springfield 1971).
Elftman, H.: The work done by muscles in running. Am. J. Physiol. *129:* 672–684 (1940).
Frigo, C.; Pedotti, A.: Determination of muscle length during locomotion; in Asmussen, Jorgensen, Biomechanics VI-A, International Series on Biomechanics, vol. 4A, pp. 355–360 (University Park Press, Baltimore 1978).
Gray, S.: Predisposing factors in thigh muscle strain in sport. XXth Wld Congr. Sports Medicine Congr. Proceedings, Melbourne 1975, pp. 325–332.
Komi, P.V.: Neuromuscular performance factors influencing force and speed production. Scand. J. Sports Sci. *1:* 2–15 (1979).
Hay, J.G.; Reid, J.G.: The anatomical and mechanical bases of human movement (Prentice-Hall, New Jersey 1982).
Kunz, H.; Kaufmann, D.A.: Biomechanical analysis of sprinting. Decathletes versus champions. Br. J. Sports Med. *15:* 177–181 (1981).
Luhtanen, R.; Komi, P.V.: Mechanical factors influencing running speed; in Assmussen, Jorgensen, Biomechanics VI-B, International Series on Biomechanics, vol. 2B, pp. 23–29 (University Park Press, Baltimore 1978).
Mann, R.: A kinetic analysis of sprinting. Med. Sci. Sport Ex. *13:* 325–328 (1981).
Mann, R.; Herman, J.: Kinematic analysis of Olympic sprint performance: men's 200 meters. Int. J. Sports Biomech. *1:* 151–162 (1985).
Mann, R.; Sprague, P.G.: A kinetic analysis of the ground leg during sprinting. Res. Q. Ex. Sport *51:* 334–348 (1980).
Mero, A.; Komi, P.V.: Effects of supramaximal velocity on biomechanical variables in sprinting. Int. J. Sports Biomech. *1:* 240–252 (1985).
Nelson, R.C.; Osterhoudt, R.G.: Effects of altered slope and speed on the biomechanics of running; in Biomechanics II. Medicine Sport, vol. 6, pp. 220–224 (Karger, Basel 1971).
Sandwich, C.M.; Pacing machine. Athletic J. *47:* 36–38 (1967).

Sinning, W.C.; Forsyth, H.L.: Lower limb actions while running at different velocities. Med. Sci. Sports *2:* 28–34 (1970).

Sprague, P.; Mann, R.V.: The effects of muscular fatigue on the kinetics of sprint running. Res. Q. Ex. Sport *54:* 60–66 (1983).

Tinning, R.; Davis, K.: The effectiveness of towing in improving sprinting speed. Aust. JHPER *79:* 19–21 (1978).

Vaughan, C.L.: Biomechanics of running gait, CRC Crit. Rev. Biomed. Engng., vol. 12, pp. 1–48 (CRC Press, Boca Raton 1984).

Wickiewicz, T.L.; Roy, R.R.; Powell, P.L.; Edgerton, V.R.: Muscle architecture of the human lower limb. Clin. Orthop. rel. Res. *179:* 275–283 (1983).

Williams, K.R.: Biomechanics of running; in Exercise and sport science reviews, vol. 13, pp. 389–441 (Franklin Institute Press, Pennsylvania 1985).

Winter, D.A.; Quanbury, A.O.; Reimer, G.D.: Instantaneous energy and power flow in gait of normals; in Komi, Biomechanics V-A, International Series on Biomechanics, vol. 4A, pp. 334–340 (University Park Press, Baltimore 1976).

Wood, G.A.: Biomechanical models for human motion analysis. Aust. JHPER *76:* 35–42 (1977).

Wood, G.A.; De La Hunty, P.L.; Cresswell, A.G.: Biomechanical effects of overspeed treadmill training on sprint running. 29th Ann. Meet. Am. Coll. Sports Medicine, Minneapolis 1982.

Wood, G.A.; Marshall, R.N.; Jennings, L.S.: Optimal requirements and injury propensity of lower limb mechanics in sprint running; in Jonsson, Biomechanics X (Human Kinetics Publishers, Illinois 1987).

Dr. G. Wood, Department of Human Movement Studies, University of Western Australia, Nedlands, W.A. 6009 (Australia)

Med. Sport Sci., vol. 25, pp. 72–85 (Karger, Basel 1987)

The Influence of the Shoe on Foot Mechanics in Running

Simon M. Luethi, Alex Stacoff

Biomechanics Laboratory, Swiss Federal Institute of Technology, ETH-Zentrum, Zürich, Switzerland

Introduction

Running means putting load on the locomotor system. If an athlete experiences pain after a race or a good practice run, then this indicates that overload has occurred and that the body has reacted to it adversely. Overload can be reduced by various factors such as the intensity and duration of activity, the style of the individual, the ground surface and the shoe used.

The ambitious will accept a reduction of intensity and duration of exercise only as the worst case solution and the type of surface is mostly determined by available conditions, i.e. too often there is no other choice than to run on asphalt. Changing the individual style, like changing from heel-strike to forefoot-strike in order to reduce impact forces, would imply a longterm adaptation process. So many runners are left to alter the last of these factors, the shoe.

The shoe is the interface between the locomotor system and the physical environment of the athlete through which all the forces acting and reacting between the lower extremities and the ground are transmitted. During these processes, that is during every foot-ground interaction, the shoe has many tasks to fullfil.

The purpose of this paper is to discuss the influence of various shoe characteristics on foot mechanics and to illustrate them with a few selected examples.

Basic Considerations

A substantial number of publications dealing with biomechanical aspects of running, running shoes and running injuries can be found in the literature. Theoretical considerations, experimental results and the methodology are thoroughly described, for example by Bates et al., [1978], Subotnick [1979], Cavanagh [1980], Cavanagh and La Fortune [1980], Nigg and Luethi [1980], Clement et al. [1981], Stacoff and Kaelin [1983], Clarke et al. [1984], Frederick et al. [1984], Luethi et al. [1984], Segesser and Stacoff [1984], and Nigg [1986].

The present paper focusses on the interaction between shoe structure and foot mechanics. In an attempt to determine the effect of shoe characteristics on the mechanics of the foot it is necessary to have a look at the general kinematics and kinetics of the foot during ground contact. About 80% of the long distance runners demonstrate a distinct heel-strike pattern, the

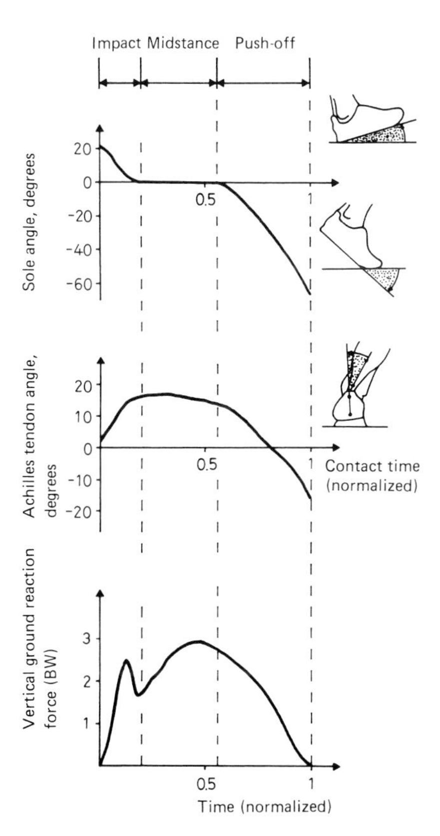

Fig. 1. The movement of the foot during ground contact (top: the angle between shoe sole and ground measured in the sagittal plane, middle: the achilles tendon angle measured in the frontal plane) and the vertical component of the ground reaction force (bottom).

remainder being midfoot and forefoot strikers [Kerr et al. 1983]. During contact the foot rotates around the principal axes of the joints of the rearfoot, that is the ankle and the subtalar joint. Figure 1 (top) shows the rolling motion of the foot in the sagittal plane represented by the angle between the shoe sole and the ground. This motion occurs at the ankle joint. The time of contact may be divided into three distinct phases: the impact phase, from the touch-down of the heel to the landing of the forefoot, the midstance phase, when both the rearfoot and the forefoot are in contact with the ground and the push-off phase, ranging from heel-off to toe-off. The movement at the subtalar joint is represented by the time history of the achilles tendon angle [Nigg et al., 1980] shown in figure 1 (middle). This variable, indicating pronation and supination, is determined by filming the test subjects from behind. Markers are placed on the lower leg and on the heel cap of the shoe or on the heel when running barefoot. Clarke et al. [1984] and Nigg [1986] have shown that the markers on the shoe move in a very similar way to those on the heel so this measurement can be used to predict the amount of pronation and supination.

Figure 1 (bottom) shows the vertical component of the external ground reaction forces over the contact time. The first pronounced force peak represents the impact of the heel and occurs within the first 30 ms of ground contact. A second peak is seen which demonstrates the reaction to the actively applied force of the runner to the ground. This force occurs during midstance and push-off phases. The meaning of the different types of forces has been widely discussed, among others, by Cavanagh et al. [1980], Nigg et al. [1980], Clarke et al. [1984], Frederick et al. [1984], Luethi et al. [1984], Denoth et al. [1985] and Nigg [1986].

The Impact Phase

As mentioned above the vast majority of long distance runners belongs to the category of heel strikers. First ground contact is usually made with the lateral aspect of the shoe sole. The touch-down velocity of the foot is mostly vertical (about 0.8–1.5 m/s) but also slightly forward (0.5–1.0 m/s). Several studies, as mentioned by Nigg [1986] show that these initial conditions are not, or only slightly affected by different types of running shoes. Immediately after first ground contact the kinematics of the foot and leg are significantly influenced by the shoe construction. The foot starts to pronate at a high rate. Maximum angular velocity of the foot relative to the leg may be as fast as 20 radians/s.

Figure 2 shows the influence of different shoe structures on the maxi-

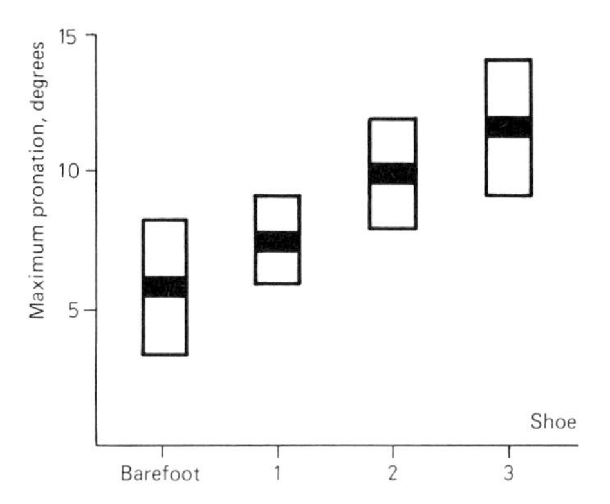

Fig. 2. Maximum amount of initial pronation for running barefoot and with three different production shoes (mean and SD for 22 subjects).

mum amount of initial pronation as compared to running barefoot. In general, a shoe increases angular displacement as well as angular velocity. This is caused by the leverage of the shoe sole with respect to the joint axes [Luethi et al., 1984; Kaelin et al., 1985; Denoth et al., 1985; Nigg, 1986]. It has been shown that the magnitude of external impact forces and the range as well as the rate of initial joint motion are interrelated [Luethi et al., 1984; Denoth et al., 1985; Kaelin et al., 1985]. Different kinematics of the foot produce different stresses on the structures of the locomotor system. If initial joint motion is kept minimal then the structures which will mainly be loaded are the soft tissue beneath the heel, the bones and the load-bearing surfaces of the joints. On the other hand, excessive initial joint motion results in fast eccentric loading of the muscles, mainly the long extensor and inverting muscles of the foot, their tendons and the supporting ligaments. Consequently, there is a high risk that the partial load on a few structures may be too heavy which can result in pain and injuries.

The purpose during the impact phase is to reduce the impact shock in general and to distribute the occurring load on different structures of the locomotor system as evenly as possible in order to avoid excessive partial load concentrations and thus decrease the risk of injuries.

The Midstance Phase

As midstance phase is defined the time over which the entire foot is on the ground. The body's center of gravity moves downward and forward over the supporting leg and rises again towards the end of this phase when the heel starts to leave the ground. This phase accounts for about 40 percent of the ground contact time. At the beginning the foot normally rolls into a pronated position, a normal functional movement determined by the anatomy of each individual foot and initiated to some extent by the initial joint motion during the impact phase. This movement takes place in all three anatomical planes. Muscular control of pronation is given by three muscles running medially underneath a structure of the calcaneus, the sustentaculum tali, which acts as a pulley for the tendons of the muscles. These three muscles are the m. tibialis posterior, the m. flexor digitorum longus and the m. flexor hallucis longus. The first two are often mentioned in connection with shin splint problems [Segesser et al., 1984]. Due to the repetitive use of these muscles at each stride, the origin of these muscles at the posterior aspect of tibia and fibula may become inflamed causing pain (shin splints). This is one of the possible reasons for this well-known running injury. The deltoid ligament which supports these muscles is located on the medial side of the rearfoot and connects the tibia to the calcaneus, talus and navicular. It is stretched in pronounced pronation.

A comparison with running barefoot shows that shoes generally increase the range of maximum pronation during the midstance phase [Nigg et al., 1980]. Favoured by a soft midsole, the shoe may force the foot into overpronation (fig. 3). According to Clement et al., [1981], overpronation is one of the most important causes for running injuries.

During midstance the ground reaction forces acting on the foot are propulsive in nature. It is therefore not the purpose to reduce the magnitude of these forces but to direct their line of action in order to control pronation.

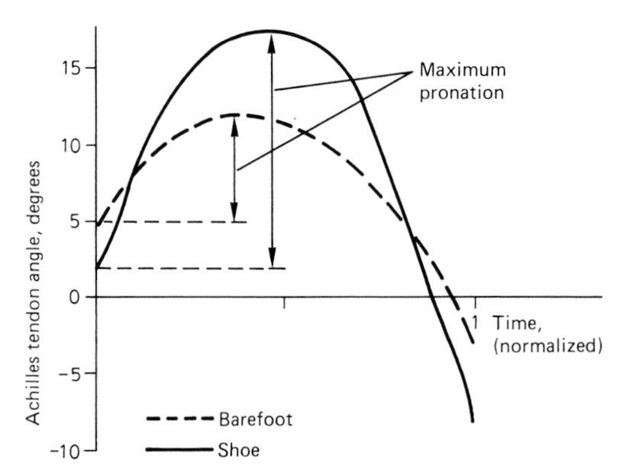

Fig. 3. Maximum amount of pronation during midstance. Comparison between barefoot and an inappropriate shoe (mean curves for 24 subjects).

The Push-Off Phase

The push-off phase begins when the heel raises from the ground and ends when the toes leave the ground. When observing a runner from behind, the take-off position of the foot relative to the ground and the leg is quite simple to recognize. There exist basically three different types of take-off positions which are shown schematically in figure 4: (a) The heel raises off the ground with no recognizable varus or valgus tilt. The rearfoot lines up more or less with the axis of the leg and the achilles tendon appears to be a straight band. (b) Possibly as a carry-over from midstance the foot remains in the overpronated position even during the push-off phase. The heel rotates internally with the rotational center in the area of the first metatarsophalangeal joint. During this type of movement the achilles tendon is curved medially (convexity to its medial border). (c) As the push-off progresses the foot supinates and the heel externally rotates. The achilles tendon is thus curved laterally.

Under barefoot conditions the foot has a straighter take-off than is generally the case when running in shoes [Nigg et al., 1980]. The calf muscles are responsible for thrusting the runner forward. When contracting, these muscles use the achilles tendon to transmit the forces onto the calcaneus. In a straight take-off the force in the achilles tendon is evenly distributed over the cross-section. The chances for any sort of inflamation produced by frictional interaction of the tendon with the peritendinous tissue are small. This is different for the runners with overpronation or oversupination during push-off. In both of these positions the force distribution in the achilles tendon is

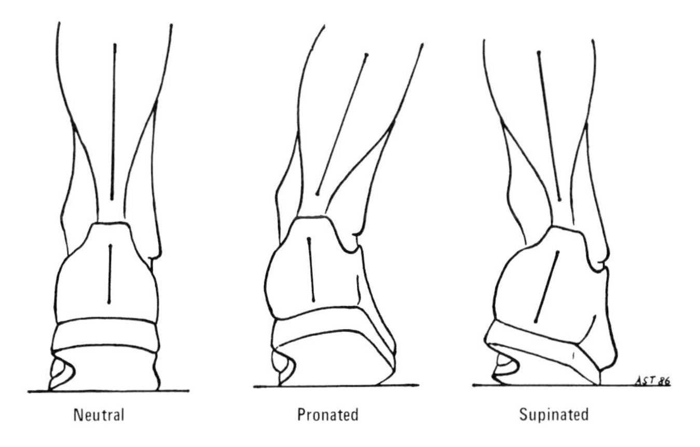

Fig. 4. The three types of push-off positions of the foot.

not homogenous. Such uneven loading leads to high force concentrations on locally very small areas within the tendon. Additionally, frictional forces between the tendon and the surrounding tissue may become significant. It is therefore not surprising that achilles tendon problems are often related, at least to some extent, to take-off overpronation or oversupination [Nigg et al., 1980; Segesser and Nigg, 1980]. A further reason for pain at the achilles tendon still stems from lax ligaments on the lateral side of the foot. Ligament laxity leads to oversupination at take-off which in turn can produce achilles tendon problems as outlined above. So the lack of function of one structure can lead to an injury of yet another structure, a chain reaction which is not easy to detect.

During this phase high force outputs are of interest since these forces determine the speed of a runner. The purpose during this phase is to control the position of the foot with respect to the leg and as a consequence to direct the acting load.

The conclusions that emerge from these basic considerations are that the running shoe has many tasks to fullfil and the following general statement may be made: with respect to the load on the different structures in the foot and the lower leg, the running shoe should *reduce, distribute* and *direct* the acting forces.

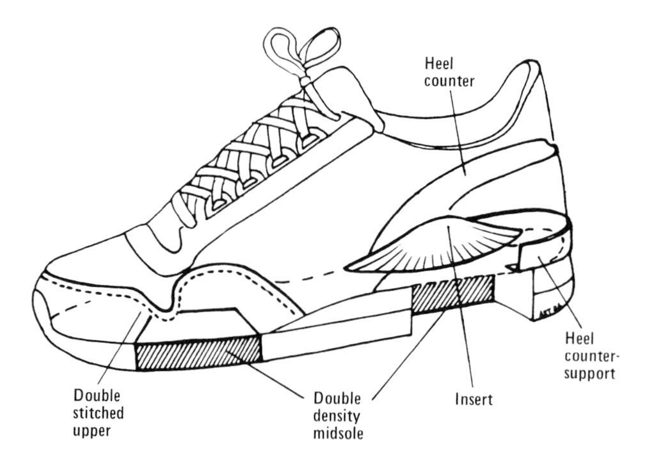

Fig. 5. Major structural components of the modern running shoe.

Consequences for the Shoe Construction

When trying to influence load distribution with the help of the running shoe there are a number of shoe elements to consider (fig. 5). Altering these elements leads to various changes of the kinematics and the load distribution in the athlete's lower extremities. Systematic testing is needed to show the influence of these elements on the runner.

Reduction and Distribution of Impact Forces

The magnitude of the occurring impact forces depends on the distance over which the different materials are deformed. Under the best condition, the ground, the shoe and the heel-pad deform together and contribute to a soft impact. However, this only works if the materials have equivalent properties. In reality this is not usually found, particularly when running on hard surfaces. So it is mostly the shoe which has to produce the largest deformation, a problem which seems relatively simple to solve but which has also lead to various solutions and controversies for a number of years. For almost a decade shoe manufacturers tried to solve this problem by making the shoes as soft as possible; but this resulted in a bottoming-out effect [Denoth et al., 1985; Nigg, 1986]. This effect occurs, when all sole material is compressed before the foot has been stopped in its downward motion. Suddenly the impact becomes very hard as the material ceases to flatten further. This is represented in figure 6 for a very soft midsole (material C). Figure 6 further

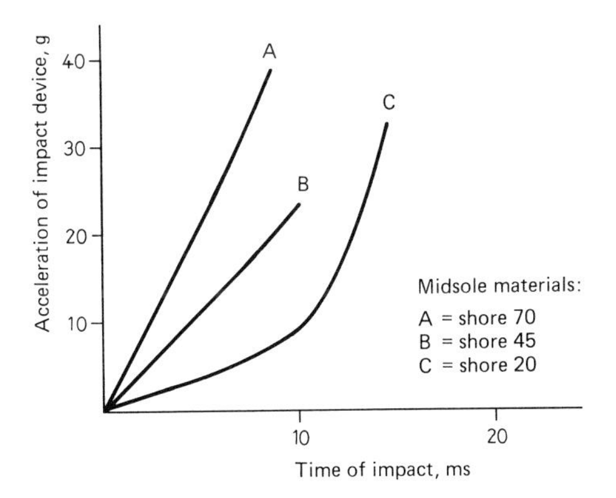

Fig. 6. Results of impact tests with different midsole materials (impact device: iron shot, m = 5 kg, r = 5 cm, v = 1 m/s). Note the bottoming-out which occurs when using material C after about 12 ms.

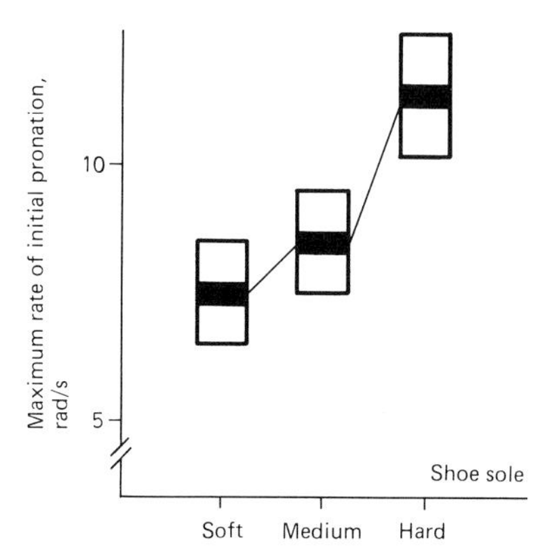

Fig. 7. The influence of shoe sole hardness on the maximum rate of initial pronation (mean and SD for 18 subjects).

shows the effect of a very hard sole (material A) on impact forces and the result for a medium hard sole (material B) which is optimally deformed and therefore demonstrates the most effective load reduction.

When trying to decrease the risk of injury, no single structure such as a tendon or ligament should carry a dominant portion of the load. The idea is to distribute this load to a number of different structures which carry it together. In this distribution the shoe can contribute significantly, particularly with respect to impact forces.

Theoretically, an increase in sole thickness should reduce impact forces. This, however, works only if during the time of impact the movement at the ankle joints is not changed, in other words if the foot is not reacting immediately to the impact. But recent studies [Luethi et al., 1984; Denoth et al., 1985; Kaelin et al., 1985; Nigg, 1986] have shown that the foot movement is actually changed during touch-down and that this change is related to the hardness and thickness of the sole material (fig. 7). As mentioned before thick and hard shoe soles have an increased lever arm about the subtalar joint axis (fig. 8). As a consequence the foot is forced into a fast initial joint motion. This means that the muscles, which should balance the moments, are eccentrically loaded at a high rate. Fast and uncontrolled movements increase the risk of injuries such as shin splints and knee injuries as mentioned before.

There are two main requirements on the shoe construction for the impact phase. Firstly, the hardness of the sole has to be chosen such that a bottoming out effect does not occur. A reasonable range for the hardness for

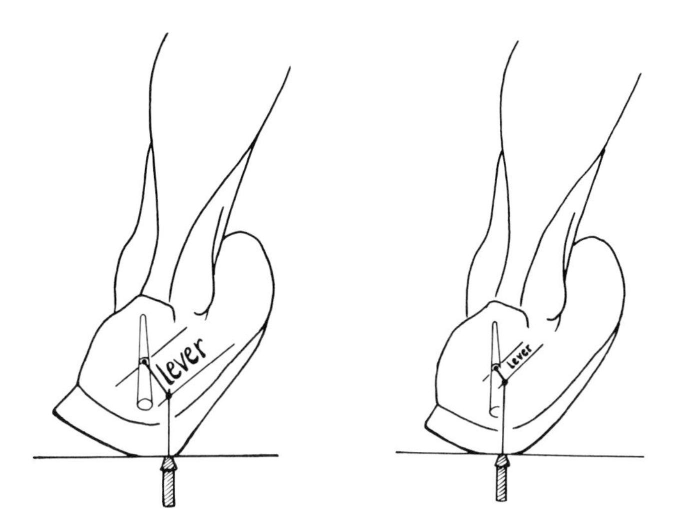

Fig. 8. The leverage of the shoe sole with respect to the subtalar joint axis.

running shoes is about 35–45 shore A (shore A is the standard measure of material hardness among shoe manufacturers) which corresponds to a medium hard shoe sole. Secondly, the thickness of the sole has to be modified to decrease the leverage about the subtalar axis. By doing this the entire geometry of the heel of the shoe is altered. The sole width becomes narrow and the heel of the sole becomes rounded (fig. 8). Recent development in industry shows the first products along this line.

Control of Pronation during Midstance Phase

To control pronation means that the foot and the shoe should work together such that no excessive pronation can occur. If the runner's anatomy cannot provide enough muscular and ligamentous support then the shoe should help to control pronation. There are a number of elements of shoe construction which can be considered to achieve this: the double density midsole, the insert, the heel counter and the heel counter supports (fig. 5).

The first two of these elements achieve the same effect. They support the foot on its medial side producing a moment opposing or controlling pronation. In other words, the sole yields less on the medial side than on the lateral side. This can be achieved by making the medial part of the shoe sole out of harder material and/or by bulding up the medial side of the sole in form of an insert. Both of these constructions are often used among manufacturers.

The heel counters and heel counter supports are hard construction elements connecting the shoe sole to the upper. They work against pronation from the side rather than from below like the sole does. Their supporting effect comes from resisting the medial movement of the foot. The pronation control of heel counters has been described previously [Stacoff et al., 1983]. The larger and stronger the heel counter, the more the pronatory movement of the foot is reduced. Figure 9 shows the effect of two different heel counters on the range of pronation where HC denotes a conventional heel cap and HCS a larger heel cap combined with a heel counter support.

One important point is to realize that putting just one or two of these elements into the shoe, does not solve the problem of overpronation. An extreme pronator will always force the weakest material to yield. In other words, if a hard insert is put into a soft shoe, then the shoe will invariably be deformed and the foot will no longer be prevented from excessive pronation. The shoe manufacturers are therefore faced with the problem of building shoes which resist overpronation equally at the sole, the insert and the heel counter.

Direction of Push-Off Forces

As stated before, neither overpronation nor oversupination is welcome during the push-off phase of the foot. It is the straight push-off which one wants to achieve and this is demonstrated only by a minority of runners. A cinematographical study conducted on 50 runners shows that those with a straight push-off accounted for only about 35% of the sample (fig. 10). The majority shows either a pronated or a supinated position of their feet.

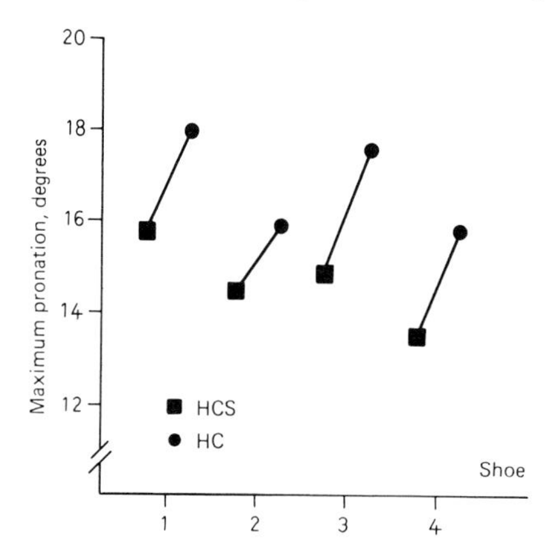

Fig. 9. The effect of two different heel counters on maximum pronation with four shoes. HC = Conventional heel cap; HCS = combined heel cap and heel counter support. n = 22.

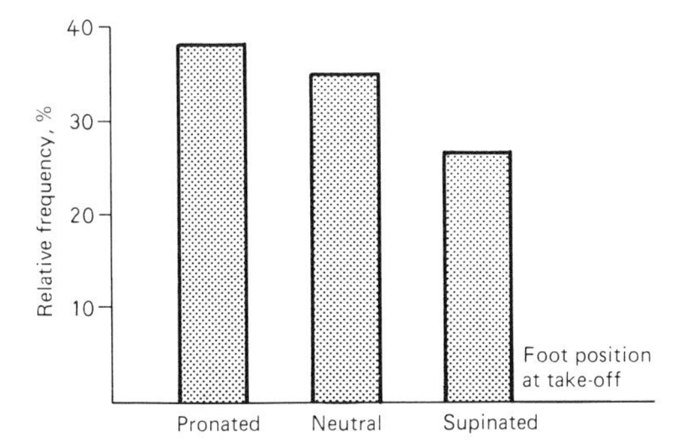

Fig. 10. Relative distribution of push-off behavior of 50 runners.

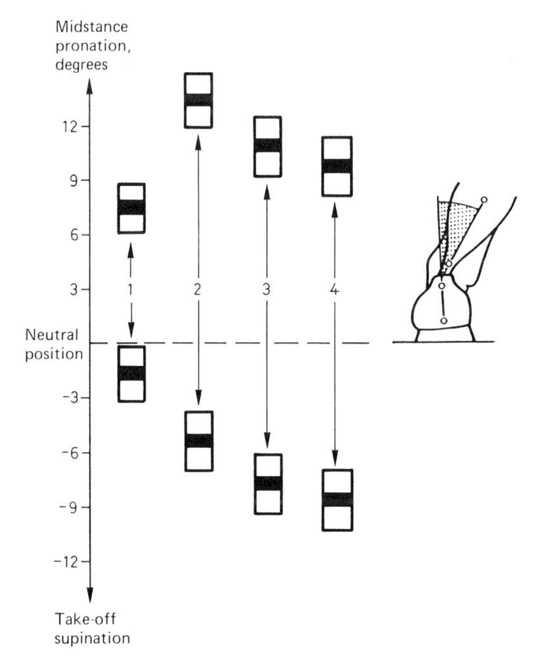

Fig. 11. The influence of a medial insert on maximum midstance pronation and take-off supination. 1 = barefoot; 2 = no insert; 3 = medial insert in the middle of the shoe; 4 = medial insert in the posterior part of the shoe. Mean and SD for 20 subjects.

The outcome for the push-off position of the foot is already prepared during the previous phases of ground contact. In other words, the preparation for a straight push-off has to be carefully balanced during the midstance phase. Figure 11 shows an example how an effective medial support may prevent excessive pronation during midstance (shoes 3 and 4), but can also lead to oversupination during push-off.

In this way a shoe element put in for one purpose may have yet another effect which should be considered when designing the shoe. Consequently, runners who are faced with an oversupination problem should look for support in the lateral forefoot. Heavy double stiched upper material and double density sole at this location of the shoe have proved to be effective in directing the foot into a straight push-off.

Concluding Critical Remarks

Research in biomechanics on running and running shoes has shown clearly that any change in a shoe has consequences for the kinematics as well as the loading of the lower extremities. In literature a number of construction strategies are proposed in order to reduce impact forces or to control overpronation and oversupination problems. Inspired by this research, but perhaps also overwhelmed by the technical possibilities the shoe manufacturers have developed a large variety of products. At the present time, functionally well-designed shoes are found on the shelves of the shoe stores right next to the latest poorly overdesigned fashion. Too often, manufactures try to build into their shoes as many ideas as possible not considering the fact that many construction elements that are good for one reason, are bad for another (i.e. shoe elements which help to reduce overpronation can enhance oversupination). This has led to some confusion on the market which does not help the runners to match their specific needs. Therefore the manufacturers should take into account that there is a need for clear distinction of different types of shoes. It is hoped that in the near future specific shoes for different requirements will be available and find their way to the runners for whom they are intended.

References

Bates, B.J.; Osternig, L.R.; Mason, B.R.: Lower extremity during the support phase of running; in Asmussen, Jorgensen, Biomechanics VI-A, pp. 31–39 (University Park Press, Baltimore 1978).

Cavanagh, P.R.: The running shoe book (Anderson World, Mountain View 1980).
Cavanagh, P.R.; Lafortune, M.A.: Ground reaction forces in distance running. J. Biomech. *13:* 397–406 (1980).
Clarke, T.E.; Frederick, E.C.; Hamil, C.: The study of rearfoot movement; in Frederick, Sport shoes and playing surfaces, pp. 166–189 (Human Kinetics Publishers, Champaign 1984).
Clement, D.B.; Taunton, J.E.; Smart, G.W.; McNicol, K.L.: A survey of overuse running injuries. Phys. Sports Med. *9:* 47–58 (1981).
Denoth, J.; Kaelin, X.; Stacoff, A.: Cushioning during running – material tests contra subject tests. Sportmedicine *7:* 197–202 (1985).
Frederick, E.C.; Clarke, T.E.; Hamill, C.: The effect of running shoe design on shock attenuation; in Frederick, sport shoes and playing surfaces, pp. 190–198 (Human Kinetics Publishers, Champaign 1984).
Kaelin, X.; Denoth, J.; Stacoff, A.; Stuessi, E.: Cushioning during running; in Perren, Schneider, Biomechanics: current interdisciplinary research, pp. 651–656 (Martinus Nijhoff, Dortrecht 1985).
Kerr, B.A.; Beauchamp, L.; Fisher, V.; Neil, R.: Footstrike patterns in distance running; in Nigg, Kerr, Biomechanical aspects of sport shoes and playing surfaces, pp. 135–142 (University Printing, Calgary 1983).
Luethi, S.M.; Nigg, B.M.; Bahlsen, H.A.: The influence of varying shoe sole stiffness on impact forces in running. 3rd Biannual Conf. Can. Soc. of Biomechanics, Human Locomotion III, 1984, pp. 65–66.
Nigg, B.M.; Luethi, S.M.: Movement analysis for running shoes. Sportwissenschaften *3:* 309–320 (1980).
Nigg, B.M.: Biomechanics of running shoes (Human Kinetics Publishers, Champaign 1986).
Segesser, B.; Stacoff, A.: The role of podiatry in sport. 1st IAAF Medical Congr., 1984, pp. 45–55.
Segesser, B.; Nigg, B.M.: Tibial insertion, tendinoses, achillodynia and damage due to overuse of the foot. Etiology, biomechanics, therapy. Orthopaede *9:* 207–214 (1980).
Stacoff, A.; Kaelin, X.: Pronation and sportshoe design; in Nigg, Kerr, Biomechanical aspects of sport shoes and playing surfaces, pp. 143–151 (University Printing, Calgary 1983).
Subotnick, S.I.: Cures for common running injuries (Anderson World, Mountain View 1979).

Dr. Simon M. Luethi, Biomechanics Laboratory, Swiss Federal Institute of Technology, ETH-Zentrum, CH-8092 Zürich (Switzerland)

Med. Sport Sci., vol. 25, pp. 86–106 (Karger, Basel 1987)

Theoretical Models and Their Application to Aerial Movement

M.R. Yeadon

Biomechanics Laboratory, University of Calgary, Canada

Introduction

Investigations of human movement may be descriptive, experimental or theoretical. A descriptive study records what happened but does not attempt to explain the results. Such a study often comprises the first step in investigating a new area since it provides basic information. This information may suggest questions such as: 'Why does this happen?' or 'What would happen if...?'.

Answers to these questions may be sought using either experimental or theoretical approaches. In both approaches the idea is to determine the result of varying just one element of the movement. In an experimental study of human movement it is not possible to have such complete control. If an athlete is required to change one aspect of technique, he may also make other changes which may have a large effect on the result. With a theoretical model of human movement this problem is overcome, since just one aspect can be varied without making any additional changes.

It is the purpose of this chapter to describe theoretical models and their application to human movement whilst airborne.

Theoretical Models

A theoretical model of human movement typically comprises a number of rigid 'segments' which are connected at 'joint centres'. The relative orientations of two adjacent segments may be described using up to three angles, and the number of degrees of freedom at the common joint centre is defined as the number of such angles. If the position of one point of the body and the orientation of one segment in space are also specified then the locations and orientations of all the segments are defined.

For a movement such as a headspring, in which all motion is parallel to a single plane, it is sufficient to use one orientation angle for each segment and two coordinates to specify the position of one point of the body. In such a two-dimensional or planar model the number of degrees of freedom will be $(n+2)$ where n is the number of segments. For a general three-dimensional model there will be three angles for each of the n segments and three coordinates for the position of one point of the body so that the total number of degrees of freedom will be $(3n+3)$.

The motion of the system of linked rigid bodies is governed by equations derived from Newton's laws of motion. These differential equations may be solved for the 'output variables' describing the motion provided that the remaining 'input variables' are specified.

Before a model is used to determine what happens in hypothetical situations it is necessary to evaluate the model using data taken from actual performances. This is done by using real values for the input variables and comparing the values of the output variables given by the model with the values derived from the performance data. In this way it is possible to obtain an estimate of the accuracy of the model.

An appropriate choice of model and associated input variables is dependent upon the type or the phase of the movement. In the take-off phase of the long jump, for example, the characteristics of the athlete's muscles limit the vertical velocity of the whole body mass centre. Thus, a comprehensive model of the long jump could include subject specific muscle models and use muscle activation levels as input variables [Hatze, 1981]. Knowledge of such input variables, however, may be difficult to obtain and this will complicate application of the model.

If the aerial phase of a gymnastic movement is of interest, the situation is quite different since it is the timing rather than the speed of limb movements that affects performance [Yeadon, 1984]. In this case orientation angles, which describe body configuration, may be used as input functions and constraints on speed of movement will be of minor importance. This procedure has the considerable advantage that, once the desired configurational changes have been visualized, it is a simple matter to specify the time histories of the orientation angles.

Models with Analytical Solutions

When the differential equations describing the motion can be solved in terms of 'well-known' functions, there is said to be an analytical solution. When this is not possible and the differential equations have to be integrated numerically, there is said to be a numerical solution. The advantage of an analytical solution is that it provides a whole family of solutions as shown in the following example.

During the aerial phase of a sports movement, the only external forces acting on the human body are those exerted by the air and by gravity. The forces due to movement relative to the air have a considerable effect in activities such as ski-jumping and sky-diving, but are of minor importance in most other aerial movements. If it is assumed that the only forces acting on the human body are due to the gravitational attraction of the planet, these forces are equivalent to a constant single vertical force (body weight) acting through the mass centre of the body. As a consequence, the mass centre experiences a constant downward acceleration and the trajectory of the centre of mass is a parabola.

The fact that the mass centre of the body experiences a constant acceleration is a general and powerful result which arises from the fact that each particle of the body experiences a gravitational force proportional to its mass.

The equation of motion may be written in the form:

$$\ddot{r} = g, \qquad (1)$$

where r is the position vector of the mass centre and g is the acceleration due to gravity.

This equation may be integrated analytically to give:

$$r = r_o + \dot{r}_o t + \tfrac{1}{2}gt^2, \qquad (2)$$

where r_o is the initial position of the mass centre, $\dot{r}_o$ is the initial velocity of the mass centre and t is the time.

Equation (2) represents all solutions for the aerial movement of the mass centre and forms a useful general result. It may be used, for example, to calculate the times spent airborne by a diver who dives from various heights. No matter what choice of model is made for aerial movement, the motion of the mass centre of the body is given by equation (2). All that remains, therefore, is to describe the rotational motion of the body. Since the external (gravitational) forces have no moment about the mass centre of the body, the angular momentum about the mass centre is constant. This principle is known as 'the conservation of angular momentum' and forms the basis for determining the rotational motion during aerial movement.

The simplest model is a single rigid body which somersaults about a horizontal axis passing through the mass centre. If the somersault angle Φ is defined as the angle swept out by any line in the body lying in the sagittal plane, the angular momentum equation may be written as:

$$h = I_{\Phi} , \tag{3}$$

where h is the angular momentum about the mass centre, I is the moment of inertia about the horizontal axis through the mass centre and ϕ is the angular velocity.

Since h and I are constants, $\dot{\phi}$ is constant and the somersault angle ϕ increases at a constant rate. Thus, the final angle of somersault will be proportional to the time spent airborne. This result may be obtained by integrating equation (3) to give:

$$\phi = ht/I, \tag{4}$$

where t is the time spent airborne.

A rigid body model may also be used to represent a twisting somersault since, during the twisting phase, the body often maintains what appears to be a fixed configuration. If the body is extended with arms adducted, it may be modelled by a rod which has the longitudinal axis as an axis of symmetry and has principal moments of inertia A, B and C where $A = B$ and $A > C$.

Let θ be the angle of tilt between the longitudinal axis L of the rod and the invariable plane P, which is normal to the angular momentum vector h (fig. 1). The rates of change $\dot{\phi}$ and $\dot{\psi}$ of the angles defining somersault and twist are [Yeadon, 1984]:

$$\dot{\phi} = h/A, \tag{5}$$
$$\dot{\psi} = (h/C - h/A)\sin\theta, \tag{6}$$

where the tilt angle θ remains constant.

Equations (5) and (6) show that the somersault and twist rates remain constant. Similar equations are given by Eaves [1969] and Frolich [1979]. The equations may be integrated to give the number of twists per somersault as:

$$\psi/\phi = (A/C-1)\sin\theta. \tag{7}$$

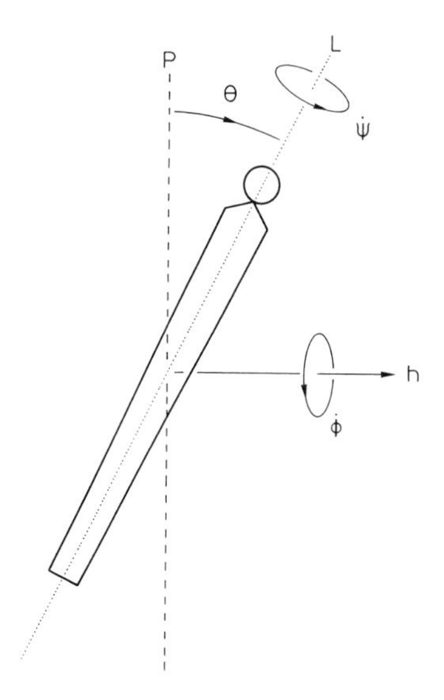

Fig. 1. A rigid body model of a twisting somersault with tilt angle θ, twist rate $\dot{\psi}$ and somersault rate $\dot{\phi}$.

Whittaker [1937] used a general rigid body model with principal moments of inertia A, B and C (A > B > C) to derive the angular momentum equations in the form:

$$\dot{\phi} = h(A\sin^2\psi + B\cos^2\psi)/AB, \quad (8)$$
$$\dot{\theta} = -h(A-B)\cos\theta\sin\psi\cos\psi/AB, \quad (9)$$
$$\dot{\psi} = (h/C - \theta)\sin\theta \quad (10).$$

These equations may be integrated in terms of Jacobian elliptic functions. There are two general solutions which may be named the *rod mode* and the *disc mode* [Yeadon, 1984]. In the rod mode, the tilt angle nutates between limiting values which have the same sign and the motion comprises a twisting somersault, with the twist angle ψ always increasing. In the disc mode, the tilt and twist angles oscillate about zero so that the motion may be described as a somersault with wobble. It is a remarkable fact that any free motion of a rigid body must fall into one of these two distinct classes. For a multi-link system such as the human body, there exists the possibility of changing from one mode to the other so that a twisting somersault could be converted into a wobbly somersault as noted by Nazarov [1978]. Such a transformation would effectively stop the twist.

If the complexity is increased by adopting a model comprising two linked rigid bodies then, in general, the equations of motion must be solved numerically and general analytical solutions are not available. However, if it is assumed that the total angular momentum is zero, there are a number of movements which may be modelled to give analytical solutions.

The following model gives an explanation of how airborne twist can be produced by relative movements of body segments. In figure 2, the human body is modelled as two identical circular cylinders with longitudinal axes U and L. During a 'hula' movement of the hips the body successively passes through: forward pike, side bend, back arch, side bend on the opposite side and forward pike again. There is a certain amount of angular momentum associated with this circling movement. In order that the total angular momentum remains zero, the whole system rotates about the vertical axis V in a direction opposite to the hula circling of the hips. The angular momentum equation takes the form:

$$I_h\dot{\psi}_h - I\dot{\psi} = 0, \tag{11}$$

where ψ_h is the angle traced out by the hips during the hula movement, I_h is an inertia term associated with the hula movement, I is the moment of inertia

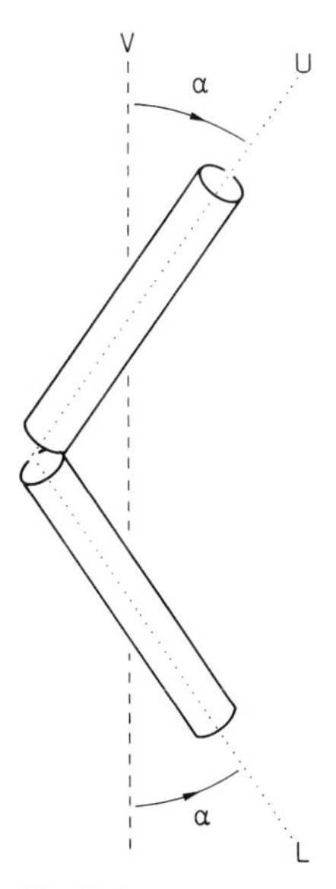

Fig. 2. A two-segment model permitting hula movement.

of the whole body about the vertical axis V and ψ is the angle through which the whole body twists about axis V.

The inertia terms I and I_h are given by:

$$I = 2(A\sin^2\alpha + C\cos^2\alpha), \quad (12)$$

$$I_h = I - 2C\cos\alpha, \quad (13)$$

where A and C are the transverse and longitudinal moments of inertia of the cylinders about their respective mass centres and α is the angle between each longitudinal axis and the vertical V.

Since I and I_h are constants, equation (11) may be integrated to give:

$$\psi = I_h\psi_h/I. \quad (14)$$

Equation (14) is equivalent to the equations derived by Kosa [1968] and Frolich [1979]. Since I_h is less than I, the twist angle ψ is always less than the hula angle ψ_h so that one cycle of hula movement will produce less than one twist. The twist angle ψ will be large when A/C and α are large. In other words, the cylinders should be long and thin and with their longitudinal axes flexed out of alignment.

This model gives a clear description of how a cat can twist when dropped upside down. Qualitative explanations which refer only to movement about the two axes U and L [Rackham, 1960; Dyson, 1973] are less convincing. Kane and Scher [1969] used an extension of the model which permitted the angle α to vary but the solution required the numerical integration of equation (11).

In each of the examples described above it was possible to obtain an analytical solution integrating the angular momentum equation. Equation (14), for example, gives the amount of twist arising from a hula movement for all such two cylinder models. Thus, an analytical solution comprises a whole class of solutions rather than one particular solution. This ability to provide a general solution describing a whole class of movements is the main strength of the analytical method.

Models with Numerical Solutions

When the differential equations of motion do not possess a general analytical solution and are solved numerically using a digital computer, the model is often referred to as a computer simulation model. Such computer models can provide only individual simulations whereas models with general analytical solutions may be considered to produce whole classes of simulations.

The main advantage that computer simulation models have over analytical models is that they can be more complex and can simulate many different kinds of movement. This versatility makes the evaluation of a model using real data a viable proposition. In order to evaluate an analytical model with a limited number of degrees of freedom, it is necessary to obtain data on performances which conform to the limited movements that can be modelled. The problem here is similar to that of an experimental study since a human subject may have difficulty in suppressing additional movements. This may explain why there have been few attempts to evaluate the accuracy of simple analytical models.

Planar models of diving [Miller, 1970] and gymnastics [Nissinen et al., 1985] have been evaluated using film data of performances. In both cases the accuracy was considered to be acceptable.

Three-dimensional models of high jumping [Dapena, 1979] and trampolining [Van Gheluwe, 1981] have also been evaluated using film data. The results of such models, whose accuracy has been evaluated, can be accepted with some confidence. The results of other models of aerial movement [Scher and Kane, 1969; Passerello and Huston, 1971; Ramey, 1973; Pike, 1980], which have not been evaluated, should be viewed with some caution.

A Computer Simulation Model

In this section a computer model is presented in some detail. The body is modelled using 11 rigid segments [Yeadon, 1984] as shown in figure 3. The following notation is used: C = chest-head; T = thorax; P = pelvis; A1 = left upper arm; A2 = left forearm-hand; B1 = right upper arm; B2 = right forearm-hand; J1 = left thigh; J2 = left calf-foot; K1 = right thigh; K2 = right calf-foot.

The 11 segments are connected at 10 joint centres and the internal configuration is specified by 14 orientation angles which define the relative orientations of the segments. A frame of reference f, which moves with the system, is used to define the whole body orientation using the angles, Φ, θ, and ψ which correspond to somersault, tilt and twist (fig. 4).

It should be noted that the number of segments of the model may be increased to 16 by introducing joint centres at the neck, wrists and ankles.

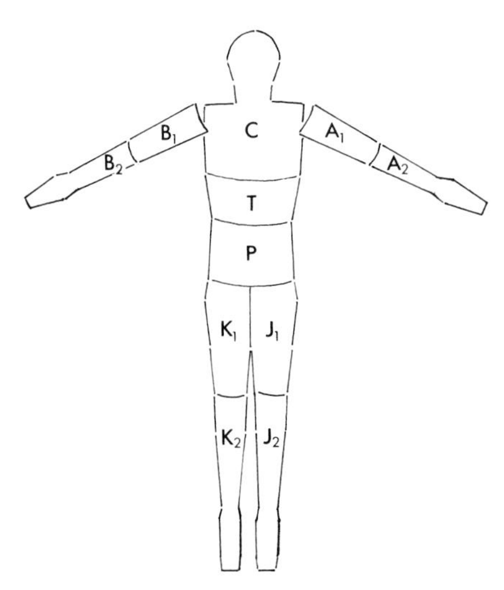

Fig. 3. An eleven segment model of the human body.

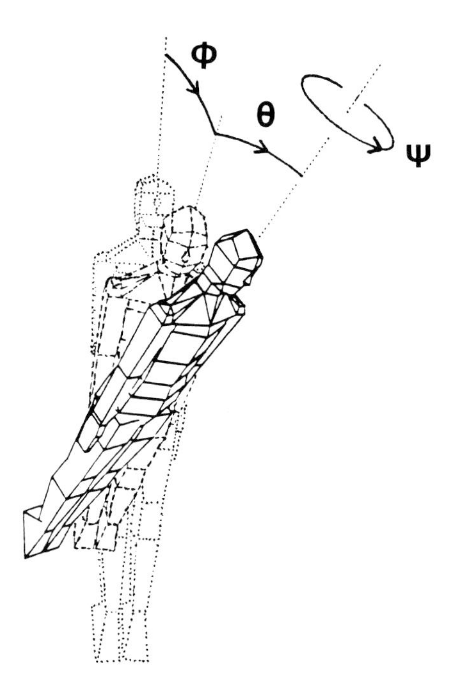

Fig. 4. Angles of somersault (ϕ), tilt (θ) and twist (ψ).

This would not necessarily be an improvement, however, since movement at these joint centres cannot be determined from film data with much accuracy.

Equation of Motion

Greenwood [1965] shows that the angular momentum of a rigid segment S about a point F is given by:

$$h_{si} = I_{ss}\omega_{si} + m_s s_f \times \overset{\cdot i}{s}_f , \qquad (15)$$

where I_{ss} = inertia tensor of S referred to its mass centre; ω_{si} = angular velocity of S relative to the non-rotating frame i; m_s = mass of S; s_f = position vector of mass centre of S relative to the point F, and $\cdot i$ = a vector time derivative in frame i.

If the point F is the mass centre of the system of 11 rigid bodies, the total angular momentum about the mass centre will be:

$$h = \Sigma\, h_{si} , \qquad (16)$$

where the sum is taken over the 11 body segments. Equation (16) may be expressed in the form:

$$h = I_{ff}\omega_{fi} + h_{rel} , \qquad (17)$$

where I_{ff} is the whole body inertia tensor, ω_{fi} is the angular velocity of the reference frame f, which rotates with the system, and h_{rel} is the angular momentum due to movements of segments relative to the reference frame f.

In a simulation, the total angular momentum h and the relative movements of body segments must be specified. Equation (17) is then solved for ω_{fi} which gives the angular rates of change $\dot{\phi}$, $\dot{\theta}$ and $\dot{\psi}$. This system of differential equations is then integrated numerically to give the time histories of the orientation angles ϕ, θ and ψ.

Evaluation of the Model

The segmental inertia parameter values for a given individual are determined using 63 anthropometric measurements as input to a mathematical model [Yeadon, 1984]. The time histories of the orientation angles in an actual movement are determined from the film taken by two cameras. To obtain a simulation of such a movement, equation (17) is used to calculate the total angular momentum h throughout the flight using the film data. Because of errors in the film data, the calculated values for h will be only approximately constant. The average of these values is used for input into the simulation model. The time histories of the 14 internal orientation angles, determined from film, are used as input data and the somersault, tilt and twist angles are given as output by the simulation. These values of the three angles may then be compared with the values obtained from film to estimate the accuracy of simulation.

The simulation model was evaluated in this way for nine twisting somersaults involving three trampolinists. The maximum deviations as percentages of total rotation were 3% for somersault and 9% for twist. These levels of accuracy should be borne in mind when drawing conclusions from the results of simulations. Figure 5 compares a graphics sequence based on a simulation with the graphics sequence obtained from film of a forward somersault with 1½ twists. The graphics sequences were produced using the SAMMIE system described in Kingsley et al., [1981].

Results Obtained Using the Models

Equation (2) describes the motion of the mass centre of the body during the flight phase of a sports movement. It may be used to find the flight times

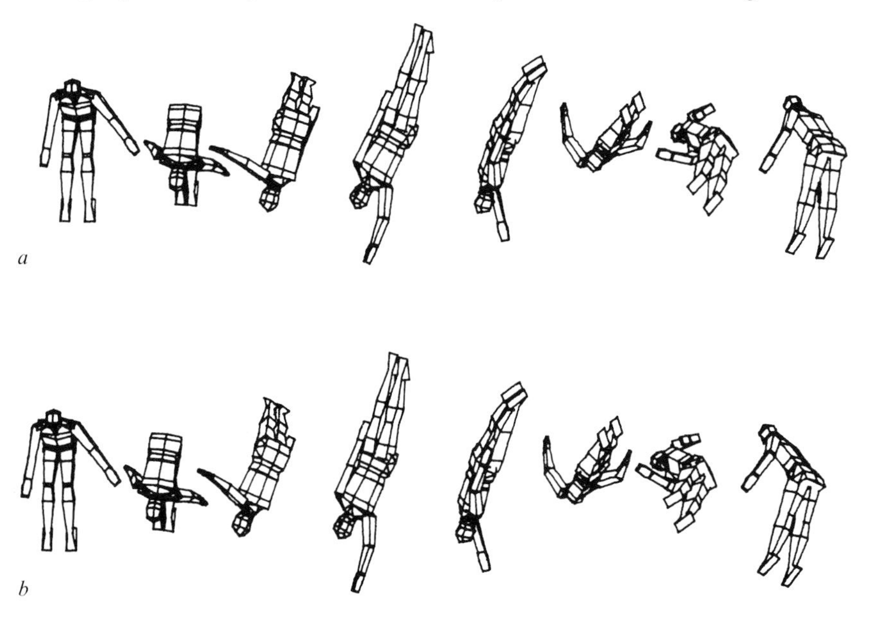

Fig. 5. Comparison of *(a)* film and *(b)* simulation.

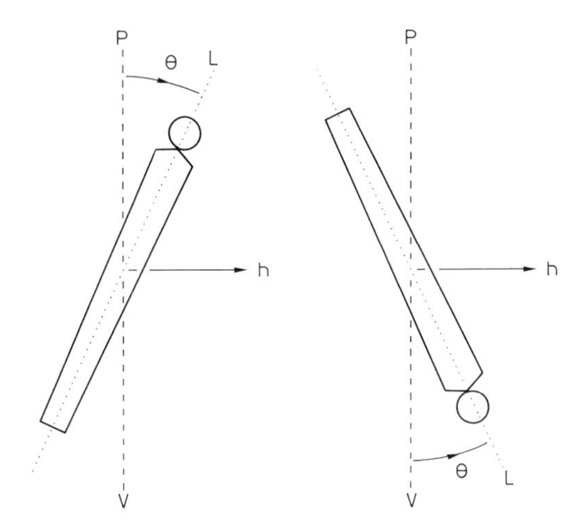

Fig. 6. In a sustained aerial twist the invariable plane P coincides with the vertical plane V.

for dives from the 1- and 3-metre springboards. If the centre of mass of the diver rises 1.5 metres before falling, the flight times are approximately 1.2 s from the 1-metre board and 1.4 s from the 3-metre board.

If a diver is to attempt 3½ sommersaults from the 3-metre springboard, the above times of flight may be used to calculate the equivalent number of somersaults from the 1-metre board. Since the divers will spend the major part of the aerial phase in a fixed configuration (tucked or piked), a rigid body model should provide a reasonable approximation for the rotational motion. Equation (4) states that the amount of somersault is proportional to the flight time. Thus, the diver should be able to perform approximately 3 somersaults from the 1-metre board.

The rod model of a twisting somersault gives the result that the longitudinal axis of the body makes a constant angle θ with the plane normal to the angular momentum vector. If the body is initially somersaulting about a horizontal axis and the tilt angle θ is produced after take-off by means of relative segmental moments, the angular momentum vector will be horizontal. In this case the invariable plane P, which is normal to the angular momentum vector h, will be coincident with the vertical plane V normal to the direction of the lateral axis at take-off. Figure 6 shows the situation after the tilt has been produced. The apparent tilt of the longitudinal axis L away from the vertical plane V remains equal to the angle θ between the longitudinal axis and the invariable plane P.

If twist is produced during the contact phase the situation is quite different. At take-off, both somersault and twist will be present and so the angular momentum vector h will not be horizontal. As a consequence, the vertical plane V will not coincide with the invariable plane P. Since the longitudinal axis makes a constant angle θ with the invariable plane P, the apparent angle of tilt, between the longitudinal axis and the vertixal plane V, will vary between zero and 2 θ (fig. 7).

The general rigid body model, with unequal principal moments of inertia, shows that the tilt angle, between the longitudinal axis and the invariable plane, oscillates between limiting values. This effect is known as nutation and is quite different to the variation in the apparent angle of tilt described above.

In constrast to this general result obtained from a rigid body model, the results obtained from computer simulation models are more specific as shown in the following examples.

Nissinen et al. [1985] simulated a hypothetical tucked dismount from the high bar, using the same initial values as for a filmed layout double backward somersault dismount. It was found that three somersaults could be completed using the tucked position.

Using the same procedure of modifying the limb movements during the flight phase, Dapena [1979] was able to show that, although a high jumper dislodged the bar set at 2.08 m, a height of 2.14 m could have been cleared.

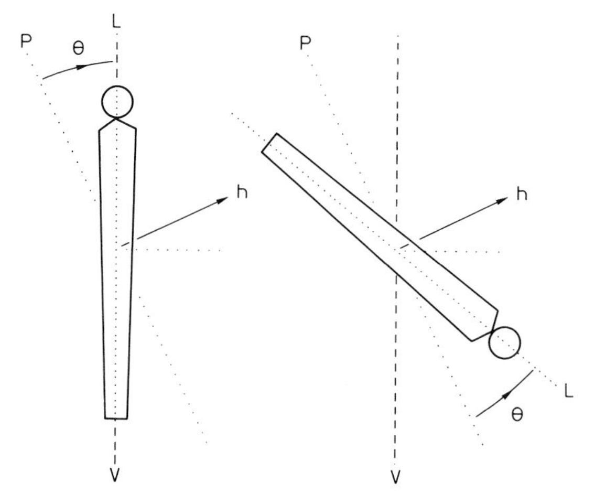

Fig. 7. Variation in the apparent angle of tilt between the longitudinal axis L and the vertical plane V when twist is present at take-off.

Scher and Kane [1969] showed that it is theoretically possible to convert a pure twist into a pure somersault by means of repeated asymmetrical arm movements; whereas Pike [1980] showed that a full twist could be introduced into a plain dive using asymmetrical arm movements.

In order to identify the segmental movements responsible for producing aerial twist in backward somersaults, Van Gheluwe [1981] modified the filmed movements by first removing asymmetrical arm movement and then removing asymmetrical hip movement. The resulting simulations showed that most of the twist resulted from asymmetrical arm movement.

Simulations of Twisting Somersaults

The remaining results on aerial movement are obtained using the 11 segment simulation model [Yeadon, 1984].

In figure 8a, both twist and somersault are present at take-off and the arms are rapidly adducted as soon as the body is airborne. With arms adducted, the body is approximately axially symmetric and the tilt angle θ

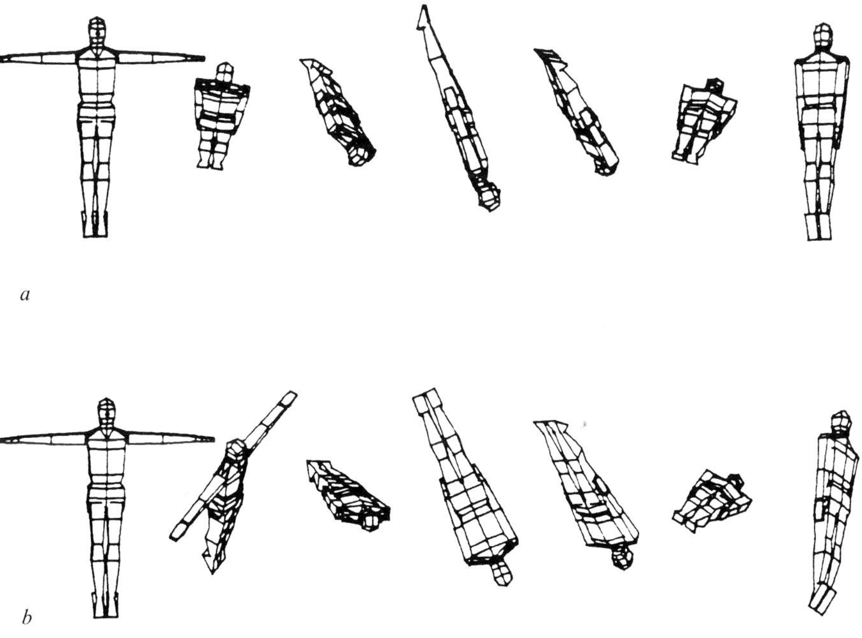

Fig. 8. The effect of arm adduction *(a)* immediately after take-off, and *(b)* at the quarter twist position.

between the longitudinal axis and the invariable plane remains approximately constant.

In figure 8b the initial conditions are exactly the same but the arms are not adducted until the quarter twist position is reached. During the first quarter twist the nutation is large and the angle θ increases to its maximum value. Once the arms are adducted, the nutation becomes small so that the tilt angle θ remains near this high value. Since the twist rate increases with θ, the final twist rate in figure 8b is greater than that in figure 8a. Even though the delay in arm adduction results in a slower twist rate initially, the increase in the tilt angle θ is sufficient to produce more twist after one somersault than early arm adduction produces.

The effects of piking during a twisting somersault can be more dramatic. In figure 9, the body is initially in the rod mode of rigid body motion so that the twist steadily increases. Subsequently, when the arms are adducted and the piked position is assumed, the motion is transformed into the disc mode and is perceived as a somersault with wobble. So long as the piked position is maintained the twist angle oscillates around the nearest multiple of a half twist and the twist has been effectively stopped. In this state, the tilt angle oscillates between positive and negative values. If extension were to be made from the pike, the twist rate may be increased, decreased or even reversed from the value prior to piking, depending on the value of the tilt angle when the extension is made.

Twist may be initiated after take-off by producing tilt when somersault is present. This may be done by making segmental movements which are not symmetrical about the sagittal plane. In figure 10 the tilt is produced by abducting the left arm during a forward somersault in the straight position. In a forward somersault this arm movement results in a twist to the left whereas, in a backward somersault, the right arm must be abducted to

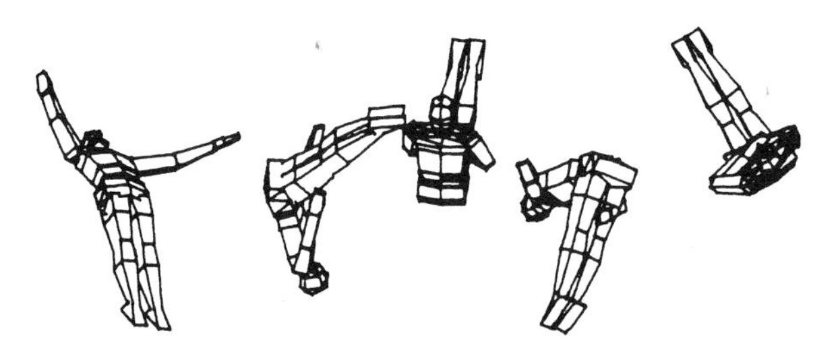

Fig. 9. Change from a twisting somersault to a wobbly somersault as a result of piking.

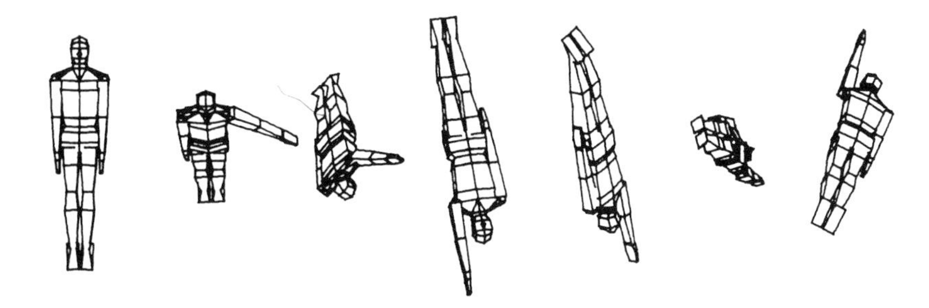

Fig. 10. Tilt produced by arm abduction.

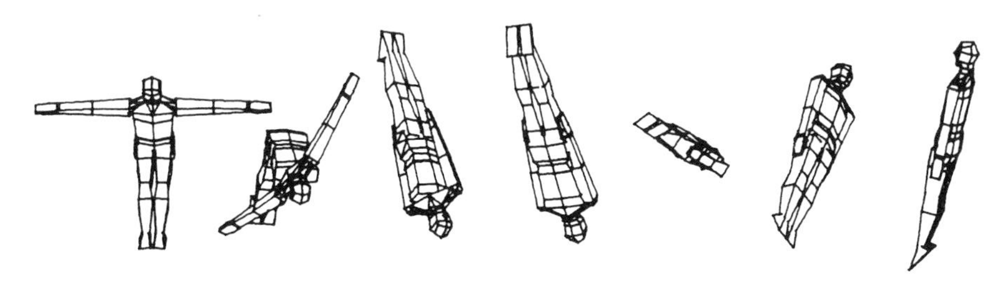

Fig. 11. Tilt produced by torsion of the chest.

produce a twist to the left. When such arm movements are made from a piked position, the resulting tilt angles and twist rates are greater.

In figure 11 tilt is produced by turning the chest to the left during a piked forward somersault. To produce a large angle of tilt, the arms should be abducted and the extension from the pike should not be made too rapidly. By delaying the extension, advantage is taken of the disc mode motion and the tilt angle rises during its oscillation cycle. In a piked backward somersault the tilt produced by turning the chest to the left results in a twist to the right whereas in a forward somersault the twist is to the left. As a consequence the tilt angle is at a different point of its oscillation cycle and the tilt becomes smaller instead of larger. The resulting twist rate upon extension from the pike is small. If the chest torsion occurs during a backward somersault when the body is arched then sufficient tilt *is* produced, the direction of twist being the same as the direction in which the chest is turned.

In figure 12 the body moves from a piked position into a side bend position during a forward somersault. As with the previous technique, in order to obtain a large angle of tilt in a piked forward somersault, it is beneficial to delay the extension from the piked position. In a piked back-

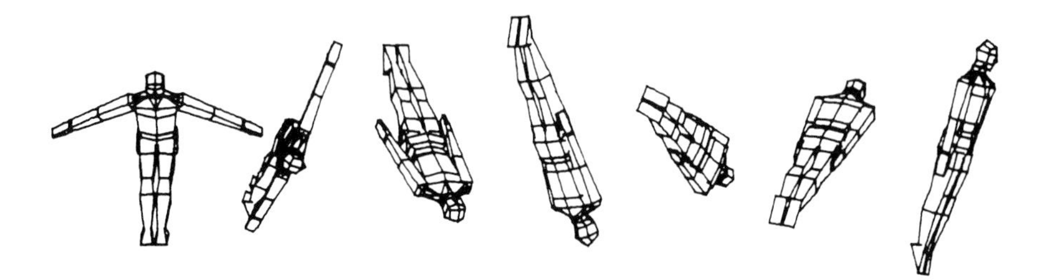

Fig. 12. Tilt produced by a partial hula movement.

ward somersault, the technique is not effective whereas, in an arched backward somersault, moving into a side pike before extending does result in a sustained twist. The twist resulting from the tilt produced by the partial hula movement is in the same direction as the twist produced by the hula movement when no somersault is present. As a consequence, there is a natural progression from twist produced by a complete hula movement in a jump to twist produced by a hula movement during a piked forward or arched backward somersault. The progression is from a rapid complete hula movement during a somersault to a slower partial hula movement. It should be recognized, however, that the two types of twist have quite different mechanics.

Thus, asymmetrical movements of the arms, chest and hips can all produce sustained aerial twist. In piked backward somersaults it should be remembered that only asymmetrical arm movements are effective. Each of the three techniques is also capable of removing the tilt and stopping the twist, but the chest torsion and hula techniques are only effective when the body is piked and the final somersault direction is backward.

Application of the Results to Coaching

It has been shown that 3½ somersaults from the 3-metre springboard are equivalent to 3 somersaults from the 1-metre springboard. This somewhat surprising result has a useful application. If a diver wishes to learn a 3½ somersault dive from the 3-metre springboard he must be able to complete 3 somersaults from the 1-metre springboard. By practising the triple somersault from the 1 metre board the diver will obtain exactly the kind of take-off needed for the 3½ somersaults from 3 metres. If progression is to be made directly from 2½ somersaults from the 1 metre board to 3½ somersaults from the 3-metre board, it is clear that the diver should be able to execute the 2½ somersaults with ease. Similar results may be obtained for platform diving. 3½ somersaults from the 10 metre platform are equivalent to 2½ somersaults from the 5-metre platform and to 1½ somersaults from the poolside. These equivalent dives from the lower levels provide useful progressions for learning dives from the higher boards.

It has been shown that a triple backward somersault from the high bar is possible using the same initial conditions as for a layout double backward somersault. Thus, in order to perform the tucked triple somersault, a gymnast can lead into the release in exactly the same way as he does for a double layout somersault. Unfortunately, many elite gymnasts are unaware of this simple fact and start to adopt a tucked position prior to releasing the bar. As a consequence, the take-off phase is changed and less angular momentum is produced so that the gymnast may have difficulty in completing three somersaults.

If twist is initiated during the contact phase then it can be beneficial to delay abduction of the arms. Although the early twist rate will be small, this is offset by the nutation effect which increases the tilt angle so that the final twist rate is large. If a rapid early arm adduction is attempted then some arm adduction may occur prior to take-off and this will tend to reduce the amount of twist produced.

The angle of apparent tilt gives an indication of the type of twist being used. When twist is initiated during the contact phase, the angle of apparent tilt is zero after an even number of half somersaults and is maximum after an odd number of half somersaults. When a sustained twist is initiated during the aerial phase by producing tilt, the angle of tilt remains constant. It is possible to use these results to determine the type of twist being used. A large tilt angle after one somersault indicates aerial twist whereas a small tilt angle after one somersault indicates contact twist.

In the backward 1½ somersault dive with 2½ twists, some twist is usually evident at take-off. Since such contact twist will tend to produce a large angle of apparent tilt after 1½ somersaults, it is necessary to remove this tilt either by piking, to take advantage of the oscillation effect in the disc mode, or by using asymmetrical arm movement. If too much contact twist is used, the diver may have difficulty in removing the tilt prior to entry. One solution to this problem is to reduce the amount of contact twist used. This is best done by modifying the twisting technique in a single somersault and noting the change in the angle of apparent tilt at entry. Another solution is to spend sufficient time in the piked position so that the tilt angle falls to near its minimum in the oscillation cycle. However, the time spent in the piked position is also dictated by the time at which the 2½ twists are completed and by the amount of somersault momentum since a near vertical entry is desired. Thus, to obtain sufficient time in the piked position, it may be necessary to change the take-off so that the initial twist and somersault rates are altered.

The oscillation effect in the disc mode can also be used to boost the tilt angle and increase the twist rate. In the ½-in 1½-out piked on trampoline, the first half twist occurs early in the first somersault. The body then pikes and subsequently extends for 1½ twists in the second somersault. By piking early so that the body is piked at the quarter twist position, the tilt angle increases until the half twist position is reached. If the body is extended soon after the half twist position, the tilt angle remains large and the motion changes to a rapid twist in the rod mode. The changeover from piking to extending should occur between the quarter and half twist positions.

Aerial twist may be produced in a somersault using asymmetrical movements of the arms, chest or hips. If the body is straight, only asymmetrical arm movement is effective in the production of the tilt which causes the twist. Thus, in a forward or inward dive with full twist, the asymmetrical arm movement should be coached. In such dives it may be helpful to initiate some twist during contact with the board but this should be kept to a minimum since it produces undesirable tilt at entry.

In a piked backward somersault, again only asymmetrical arm movement is effective and should be used in movements such as back-in full-out piked where the twist occurs in the second somersault.

In piked forward and arched backward somersaults, all three asymmetry techniques are capable of producing rapid twisting. The particular technique or combination of techniques adopted is largely a matter of personal preference. However, in the case of a performer who has difficulty in producing aerial twist, the coach has three possible approaches available.

In order to remove the tilt and stop the twist, asymmetrical segmental movements may again be used. If the final somersaulting direction is forward, as in twisting dives, only asymmetrical arm movement is effective. If the final somersault direction is backward, as in trampoline movements, all three asymmetry techniques are capable of removing the tilt. However, the arm movement is easier to coach than the chest or hip movements.

The twisting techniques described above have been successfully applied to the coaching of twisting somersaults, both in trampolining and in springboard diving. The 11 segment simulation model has already been used to determine the contributions made by the various twisting techniques to elite performances [Yeadon, 1984]. By modifying the segmental movements of filmed performances and observing the effects on the simulated motion, it will soon be possible to give advice on how to improve even the best twisting performances.

References

Dapena, J.: A simulation method for predicting the effects of modifications in human airborne movements; PhD diss. (University of Iowa, Ames 1979).

Dyson, G.H.G.: The mechanics of athletics; 6th ed., p. 106 (University of London, London 1973).

Eaves, G.: Diving: the mechanics of springboard and firmboard techniques, p. 76 (Kaye & Ward, London 1969).

Frolich, C.: Do springboard divers violate angular momentum conservation? Am. J. Phys. *47:* 583–593 (1979).

Greenwood, D.T.: Principles of dynamics, p. 375 (Prentice Hall, Englewood Cliffs 1965).

Hatze, H.: A comprehensive model for human motion simulation and its application to the take-off phase of the long jump. J. Biomech. *14:* 135–142 (1981).

Kane, T.R.; Scher, M.P.: A dynamical explanation of the falling cat phenomenon. Int. J. Solids Struct. *5:* 663–670 (1969).

Kingsley, E.C.; Schofield, N.A.; Case, K.: Sammie. A computer aid for man machine modelling. Computer Graphics *15:* 163–169 (1981).

Kosa, F.: Mechanical study of twisting motion in mid-air. Bulletin of Institute of Sport Science, Tokyo University of Education *6:* 69–75 (1968).

Miller, D.I.: A computer simulation model of the airborne phase of diving; PhD diss. (Pennsylvania State University, University Park, 1970).

Nazarov, V.T.: Analysis of the sportsman's movements in out-of-contact state; in Landry, Orban, Biomechanics of sport and kinanthropometry, pp. 257–262 (Symposia Specialists, Miami 1978).

Nissinen, M.; Preiss, R.; Bruggemann, P.: Simulation of human airborne movements on the horizontal bar; in winter, Norman, Wells, Hayes, Patla, 9th Int. Congr. Biomechanics, Waterloo 1983. Biomechanics IX-B, pp. 373–376 (Human Kinetics, Champaign 1985).

Passerello, C.E.; Huston, R.L.: Human attitude control. J. Biomech. *4:* 95–102 (1971).

Pike, N.L.: Computer simulation of a forward, full twisting dive in a layout position. PhD diss. (Pennsylvania State University, University Park 1980).

Rackham, G.: The origin of twist. Swimming Times *37:* 116–118 (1960).

Ramey, M.R.: Significance of angular momentum in long jumping. Res. Q. *44:* 488–497 (1973).

Scher, M.P.; Kane, T.R.: Alteration of the state of motion of a human being in free fall. Technical Report 198. NASA-CR 108938 (Division of Applied Mechanics, Stanford University, Stanford 1969).

Van Gheluwe, B.: A biomechanical simulation model for airborne twist in backward somersaults. J. Hum. Mov. Stud. *3:* 5–20 (1981).

Whittaker, E.T.: A treatise on the analytical dynamics of particles and rigid bodies, p. 145 (Cambridge University Press, London 1937).
Yeadon, M.R.: The mechanics of twisting somersaults; PhD diss. (Loughborough University of Technology, Loughborough 1984).

Dr. M.R. Yeadon, Biomechanics Laboratory, Faculty of Physical Education,
University of Calgary, 2500 University Drive N.W., Calgary, Alberta T2N 1N4 (Canada)

Med. Sport Sci., vol. 25, pp. 107–141 (Karger, Basel 1987)

Physical Models in Sports Biomechanics

W. L. Bauer

University of Bremen, FRG

Introduction

The human mind often has difficulty in accepting completely abstract ideas, particularly if verbally presented. It finds it much easier to grasp concepts that are illustrated by appropriate analogies or derived from concrete examples. Figures drawn on paper, architectural plans, sketches, mathematical equations or physical models provide such examples [Cundy and Rollett, 1961]. In biomechanics, in recent years, the use of models has greatly increased. This reflects a growing appreciation that the design of mathematical models is necessary to understand the complex dynamic behavior of biomechanical systems through mathematical treatment and computer simulation.

However, the understanding of mathematical models is limited by the rapidly increasing complexity of the mathematical equations required to describe the human biomechanical system and by the computational effort involved in setting up and simulating the mathematical equations [Bauer, 1983].

When communicating with simulated systems nontrivial technical difficulties are inherent in trying to represent in the two-dimensional space of the CRT screen the tactile and visual feedback signals with which man feels most at home. Because in general the human operator works best when controlling interactively in a closed-loop, error correcting manner, his model of the system should be able to provide him with the necessary feedback.

Some of the above disadvantages may be eliminated by making use of three-dimensional solid models with working mechanisms that match the phenomenon of interest. Such models, rather understandably, are called mechanical or physical models. A physical model can be a reduced size copy of the original – equally or less detailed – or an analogy with equivalent behavior.

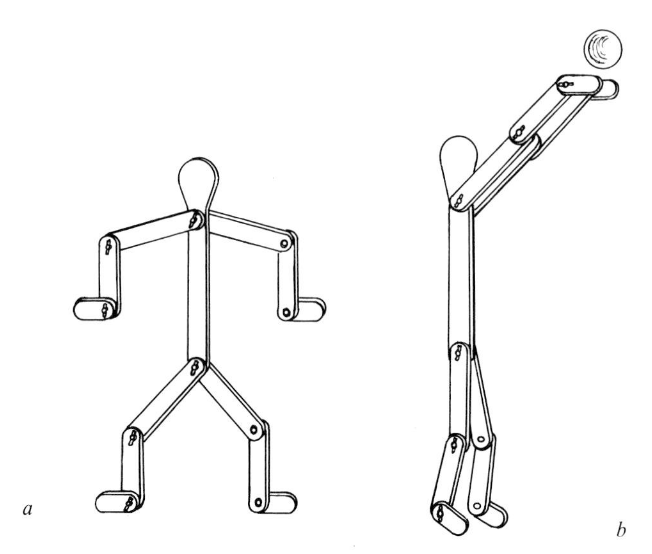

Fig. 1. Thirteen-link stick model [Balster, 1982].

Many physical models described in text books or in manufacturers' catalogues are suitable for demonstrating biomechanical principles in sport. In biomechanics of sport the mechanical part of the human body is very often modelled by plane or three-dimensional multi-link models. These models can be designed to demonstrate either static or dynamic behavior, and to illustrate all their temporal, spatial and kinetic components.

In contrast to the physical models which are made of solid objects, the mathematical models consist of a set of equations which describe relationships between physical variables of the biomechanical system under investigation. These equations can be algebraic equations, differential equations or systems of differential equations. Algebraic equations describe the relationships between the variables of interest only for steady-state behavior, they represent a static model. Differential equations and systems of differential equations represent a dynamic model which includes the steady-state relationship [Fasol and Jörgl, 1980].

Selected Physical Models

The development of reliable physical models, their design and their manufacture requires time, know-how and money. That explains why few

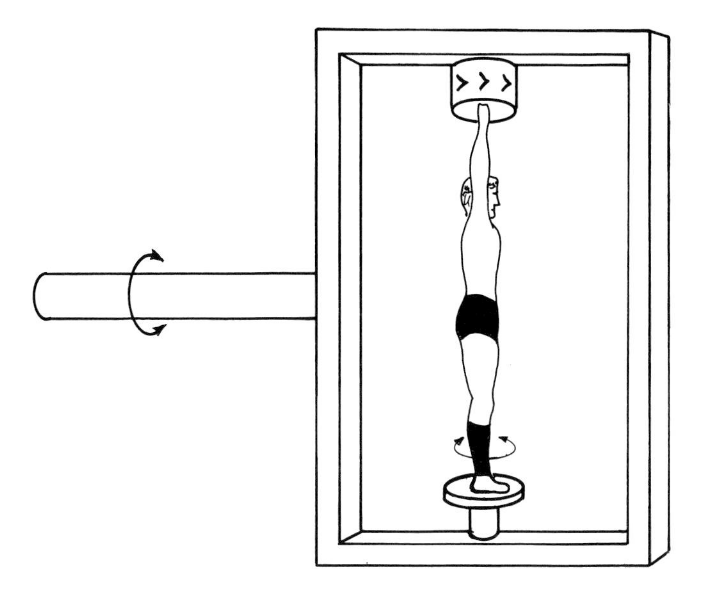

Fig. 2. Cardanic pivoted model [Nickel, 1983].

papers on that subject can be found. Publications of complete and well-tested biomechanical models, particularly dynamic models, are rare. Some of them will be presented on the following pages.

Static Models

Example 1: Model for demonstrating spatial configurations of human movements. The wooden stick model of Balster [1982] is a representative of the static model category (fig. 1a, b). The model is made of 13 plane links hinged together with friction joints. Wing screws hold the parts together and permit the variation of frictional forces within the joint. After loosening the wing screws, the model's limbs can be re-arranged into different positions, for instance into different configurations characteristic for selected motions in sport, and then frozen in these positions by re-tightening the wing screws.

The standing height of the model is 134 cm. According to the author it was successfully used to accelerate the learning process of sports movements at school during lessons in physical education.

Example 2: Model for Explaining the Importance of a Fixed Body Coordinate System for Turning and Twisting in Gymnastic Maneuvers. Nickel [1983] developed a model to justify the introduction of a reference coordinate

Fig. 3. Cardboard model for teaching the Eskimo roll [Gercken, 1984].

system fixed to the human body. It was used to denote frontal, transverse and vertical directions with respect to the upright stance and to define the sense of rotation in gymnastic maneuvers (fig. 2).

The model of the gymnast is pivoted in an erect position with arms over his head within a rectangular frame. Rotations around the vertical axis of the gymnast are to be carried out within the frame, whilst rotation around the transverse and frontal axis are to be carried out by turning the frame by hand.

Example 3: Model for Teaching the Eskimo Roll. The cardboard model of figure 3 has been used by Gercken [1984] during kayaking lessons for several years to clarify the three dimensional complex movement of the Eskimo roll and to influence the learning process positively.

Example 4: Three-Dimensional 15-Link Model of the Human Body. A three-dimensional 15-link model made of wood was used by Frohlich [1980] to illustrate and clarify the physics of somersaults and twists (fig. 4a) and from Boerchers [1986] to explain and instruct students in the complex movement of the Eskimo roll (fig. 4b).

a

b

Fig. 4. Three-dimensional fifteen-link model, respectively from *(a)* Frohlich [1980] and *(b)* Boerchers [1986].

The 15-link model is commercially available in different sizes and is very often used by art students during drawing lessons.

Dynamic Models

Example 5: Electronically Controlled Two-Link Model of a Gymnast on a High Bar. An electronically controlled two-link model of a gymnast on a

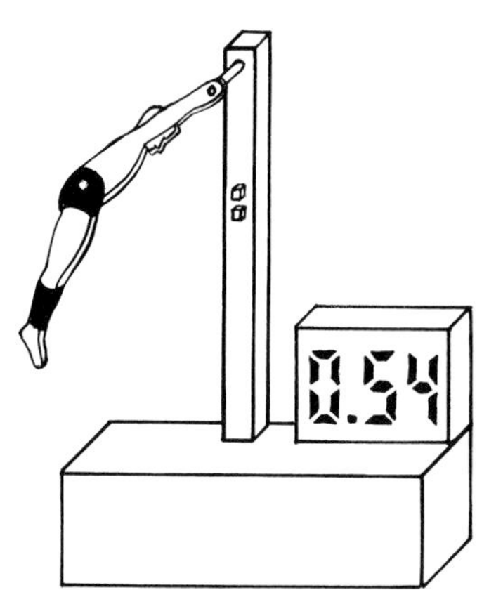

Fig. 5. Electronically controlled two-link model [Nickel, 1983].

high bar (fig. 5) was introduced by Nickel [1983]. His model has been used for demonstrating both the variations in the period of oscillation which follow from changes in hip joint configuration and also the increases in internal energy which are attendant upon improvements in the timing of those changes.

In the model, hip joint movements are controlled by an on-off electric miniature motor built into the model and powered by an NC battery. The period of the hang-swing movement is measured in microseconds by an opto-electronic counter. This device is activated on the forward swing and stopped on the following forward swing. Its time value is indicated on a seven-segment display.

The length of the gymnast model measures 45 cm from the tips of the fingers to the tips of the toes.

Example 6: Model for Demonstrating Momentum Partitioning in Diving Twisting Skills. The model, which is illustrated in figure 6, has been described by Lephart, University of Melbourne [1984; personal commun.] as follows:

Two crossed links with a pivot bolt at their intersection are free to rotate relative to each other. A rubberband is attached from one arm of the model to the upper end of its body. A string is passed from the same arm through a hole drilled in the body at the center of gravity of the system (fig. 6a).

When the model is held on its side by holding the string the rubberband is stretched and arms rotate about 60° (fig. 6b). The model is set into rotation with angular momentum H in a horizontal plane about the axis formed by

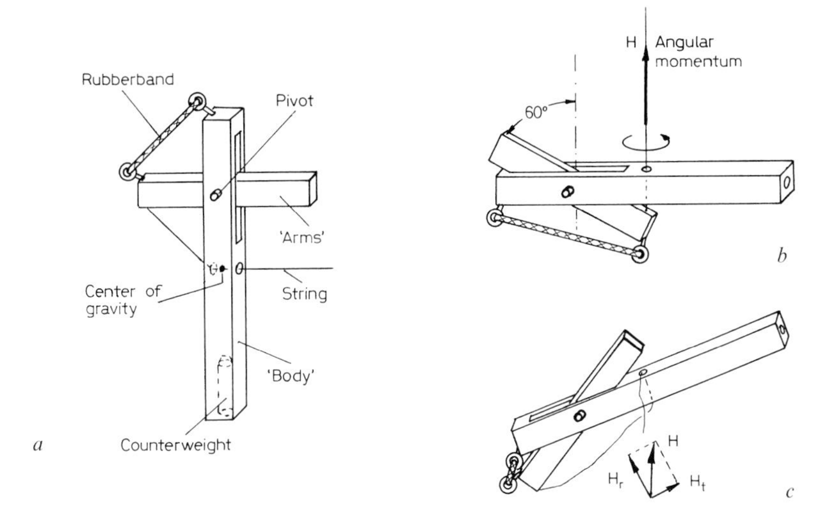

Fig. 6. Model for demonstrating momentum partitioning [Lephart, 1984].

the string. When the string is released, the model is free to fall, the rubberband causes the arms to rotate about 150°. This causes the body to twist around the longitudinal axis with the angular momentum H_t as well as to continue to rotate with the angular momentum H_r (fig. 6c).

How to Build a Physical Model

Physical models may be developed according to the following steps: (1) establish the theoretical considerations and set up the mathematical equations describing the model's dynamics; (2) observe and experiment with the real system; (3) design and manufacture a prototype model; (4) experiment with the prototype and compare the results with real system's behavior; (5) correct the model parameters and improve the model's mechanics.

Usually, the various steps must be repeated several times until the physical model achieves a suitable standard.

On a theoretical level the modelling of the dynamic properties of the system is achieved by setting up differential equations. To carry out this process correctly, and to introduce into them the appropriate laws of conservation of energy and momentum, a profound knowledge of physics and mechanics is required.

When modelling the human body by a multi-link system the mathematical model usually consists of a set of partial nonlinear differential equations with a set of algebraic equations representing the boundary conditions.

Empirical observations of the real system are made using cinematographic, dynamographic and electromyographic measurement techniques in order to obtain relevant input, output and internal state data.

These measurements are then evaluated and matched with variables of the mathematical model as, for instance, geometrical dimensions of the body segments, spatial coordinates of the segment center of gravity and values of forces applied to the apparatus in use.

At this stage the question might well be raised as to why any experimental analysis at all is required in the development of a theoretical model. The truth is that even the theoretician must make many simplifying assumptions when first formulating his equations. Consequently he must continually check these against reality by comparing the output parameters of his theoretical system with both the static and dynamic features of the real system and of its several sub-systems.

It is probably not very instructive to talk in general terms about the further developmental steps of a physical biomechanical model. The procedure will be demonstrated using the example of a mechanical gymnast on a high bar, a physical model which was presented in 1983 for the first time [Bauer, 1985].

Construction of a Mechanical Gymnast on a High Bar

Theoretical Considerations

Because of the complexity of the body and its motor system it is impossible to construct a physical model which can perform all the maneuvers of which the athlete is capable.

In the simple case of a plane motion on a high bar – defined as those movements in which the gymnast keeps both hands in the same place on the bar throughout the movement – the body may be modelled by the multi-link system shown in figure 7. The number of links may be selected according to the maneuver under investigation.

The model in figure 7a is a rigid physical pendulum which can be used to calculate approximately the acting forces, torques and the effect of different masses during giant swings [Schmidt, 1961]. To simulate the giant swing movement, the pendulum's rotation must be started in the upside down

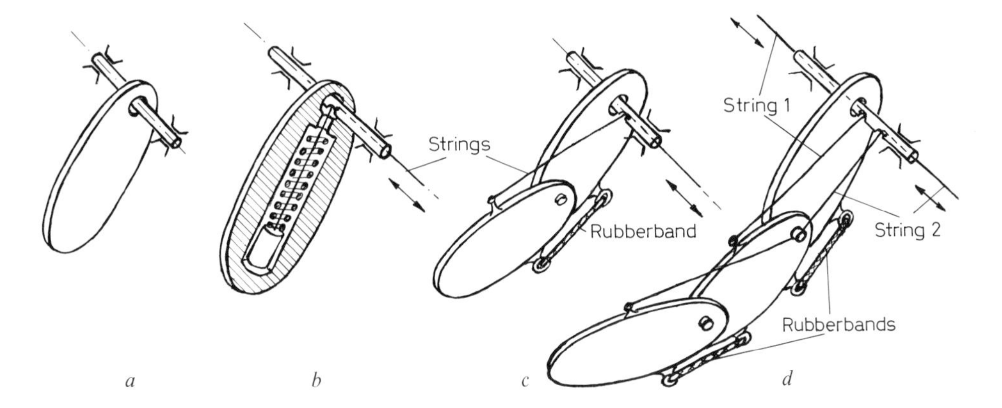

Fig. 7. Development steps for a multi-link model. *a* Simple pendulum. *b* Controllable one-link pendulum. *c* Controllable double pendulum. *d* Three-link pendulum with two controlled variables.

position with sufficient angular momentum for at least one complete revolution.

The model in figure 7b consists of a one-link pendulum with an additional mass able to slide up and down on a straight path within the system. The mass is forced to the outermost position by a spring, even when the pendulum is upside down. A string, attached to the mass, is led through a hole into, and immediately around a diversion pin, along the inside of the horizontal tubular bar. It finally emerges from the end of the bar. Pulling on the string moves the mass. In this position the lever arm of the string force on the model is effectively zero relative to the horizontal axis of rotation. Thus, no immediate torque will be exerted on the bar from the string attachment. The one-link pendulum with an additional mass behaves like an internally energized self-oscillatory mechanical system when controlled by the string from outside. However, it cannot be started from rest by simply moving the mass, that is the center of mass of the system, up and down. An exterior impulse must be applied to start the oscillation or rotation.

Once an oscillation or rotation is initiated the increase, constancy or decrease of internal energy can be demonstrated through appropriate control actions at the string arrangement which changes the center of gravity relative to the point of support by moving the interior mass up and down.

The model must be appropriately equipped to demonstrate its dynamic behavior and to indicate the forces and torques acting during a giant swing as a function of mass, angular velocity, moment of inertia, friction and of its control strategies.

The double pendulum model in figure 7c is even more realistic for such events as the giant swing. Its dynamic behavior differs from that of the model in figure 7b in that it enables the oscillations to be started from rest by appropriate controlling actions. It can be started by a rhythmic to-and-fro movement of the second link which causes positive and negative torques around the horizontal bar and rythmic changes of the center of gravity relative to the point of support. This can be done by pulling and releasing the string which is attached to the second link and comes out of the bar at one side. Pulling the string stretches the rubberband, and the second link rotates up to a certain degree. When releasing the string the rubberband causes the link to rotate back into its previous position.

Although the double pendulum is a very simple mechanical model of the human body it can give much insight into the dynamics and control requirements of a gymnast doing the forward and backward giant swing [Bauer, 1983].

A further refinement is necessary when the model is to perform the forward swing kip, the drop kip, the seat circle forward or backward, the sole circle forward or backward and the inverted hang swing. The three-link model of figure 7d will fulfill these requirements.

Since we have only three rigid links hinged together at two points representing the shoulder joint and the hip joint of the human body, stunts which require bent knees and elbows cannot be executed. Hip and shoulder joints must be controlled by voluntary muscle torques which can be chosen arbitrarily within certain limits. Compared to the previously discussed two-link model the number of manipulated variables have doubled and make the system much more difficult to control because two stimuli are simultaniously applied at different points of the system.

The two torques (stimuli) can independently be applied by two strings coming out on each side of the horizontal bar. String 1 attached to the second link can be used to bend the shoulder joint and string 2, which is attached to the third link, can be used to apply a torque to the hip joint. String 2 is passed around the axle of link 2 and link 1 in order not to change the shoulder torque during hip motions. The string arrangement only allows moments to be applied in one direction. The rubberbands have been introduced to supplement these, by invoking their reverse moments.

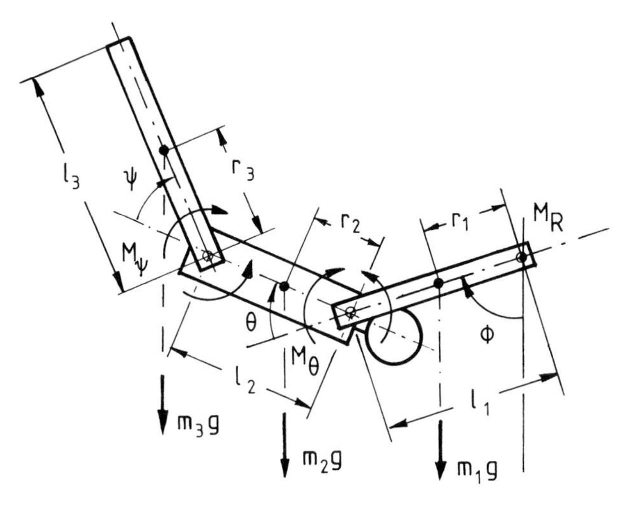

Fig. 8. Variables of the three-link model identified for mathematical description (see text).

Model Design Using Equations of Motion

Figure 8 had been used to set up the three equations of motion (1) to (3) below. These equations are derived using the Lagrange algorithm taking in account the influence of the different joint torques.

$$
\begin{aligned}
M_1 - M_R = \; & \ddot{\phi}\; [2A + 2B\cos\theta + 2C\cos(\theta + \psi) + 2D\cos\psi] \\
& + \ddot{\theta}\; [2E + B\cos\theta + C\cos(\theta + \psi) + 2D\cos\psi] \\
& + \ddot{\psi}\; [F + C\cos(\theta + \psi) + D\cos\psi] \\
& + \dot{\phi}\dot{\theta}\, [-2B\sin\theta - 2C\sin(\theta + \psi)] \\
& + \dot{\phi}\dot{\psi}\, [-2C\sin(\theta + \psi) - 2D\sin\psi] \\
& + \dot{\theta}\dot{\psi}\, [-2C\sin(\theta + \psi) - 2D\sin\psi] \\
& + \dot{\theta}^2\, [-B\sin\theta - C\sin(\theta + \psi)] \\
& + \dot{\psi}^2\, [-C\sin(\theta + \psi) - D\sin\psi]
\end{aligned} \qquad (1)
$$

where $M_1 = -[m_1r_1 + (m_2 + m_3)\, l_1]g\sin\phi + M_2$.

$$
\begin{aligned}
M_2 + M_\theta = \; & \ddot{\phi}\; [2E + B\cos\theta + C\cos(\theta + \psi) + 2D\cos\psi] \\
& + \ddot{\theta}\; [2E + 2D\cos\psi] \\
& + \ddot{\psi}\; [F + D\cos\psi] \\
& + \dot{\phi}\dot{\psi}\, [-2D\sin\psi] \\
& + \dot{\theta}\dot{\psi}\, [-2D\sin\psi] \\
& + \dot{\phi}^2\, [B\sin\theta + C\sin(\theta + \psi)] \\
& + \dot{\psi}^2\, [-D\sin\psi]
\end{aligned} \qquad (2)
$$

where $M_2 = -(m_2r_2 + m_3l_2)g\sin(\phi + \theta) + M_3$.

$$\begin{aligned} M_3 + M_\psi = \; & \ddot{\phi}\ [F + C\cos(\theta + \psi) + D\cos\psi] \\ & + \ddot{\theta}\ [F + D\cos\psi] \\ & + \ddot{\psi}\ [F] \\ & + \dot{\phi}\dot{\theta}\ [2D\sin\psi] \\ & + \dot{\phi}^2\ [C\sin(\theta + \psi) + D\sin\psi] \\ & + \dot{\theta}^2\ [D\sin\psi] \end{aligned} \tag{3}$$

where $M_3 = - m_3 r_3 g \sin(\theta + \phi + \psi)$, with

$$\begin{aligned} A &= \frac{1}{2}(I_1 + m_1 r_1^2 + I_2 + m_2 r_2^2 + m_2 l_1^2 + I_3 + m_3 r_3^2 + m_3 l_1^2 + m_3 l_2^2) \\ B &= m_2 r_2 l_1 + m_3 l_1 l_2 \\ C &= m_3 r_3 l_1 \qquad\qquad F = I_3 + m_3 r_3^2 \\ D &= m_3 r_3 l_2 \\ E &= \frac{1}{2}(I_2 + m_2 r_2^2 + I_3 + m_3 r_3^2 + m_3 l_2^2)\ . \end{aligned}$$

This system of nonlinear equations cannot be integrated without knowledge of the voluntary joint torques M_ψ and M_θ and the frictional torque M_R. Thus the equations of motion can only be solved with additional mathematical manipulations (e.g. optimization techniques).

Step 1: Approximate calculation of the maximal force exerted on the high bar. If it is assumed that the joints of the shoulder and the hip are stiff and that the model is free to swing with straight body on a smooth bar, then the following expressions within equations (1) to (3):

$$M_R, \theta, \dot{\theta}, \ddot{\theta}, \psi, \dot{\psi}, \ddot{\psi} = 0,$$

may be set to zero as indicated, where $\phi, \dot{\phi}, \ddot{\phi} \neq 0$. Thus, equation (1) simplifies as expected to the equation of a one-link physical pendulum:

$$I\ddot{\phi} + mgr\sin\phi = 0 \text{ with } I = I_r + mr^2, \tag{4}$$

where

$$\begin{aligned} I &= I_1 + I_2 + I_3 + m_1 r_1^2 + m_2(r_2 + l_1)^2 + m_3(r_3 + l_1 + l_2)^2 \\ r &= \frac{m_1 r_1 + m_2(r_2 + l_1) + m_3(r_3 + l_1 + l_2)}{m} \\ m &= m_1 + m_2 + m_3, \end{aligned} \tag{5}$$

in which I is equal to the moment of inertia with respect to the bar and r the distance of the center of gravity from the bar.

To calculate approximately the maximal radial force on the bar the equation of conservation of energy is applied. Assuming the gymnast starts in the handstand position on the bar ($\dot{\phi} \cong 0$, $\phi = 180°$, $E_p = 2$ mgr) and rotates around the bar without frictional losses ($M_R = 0$), his maximum angular acceleration is described by the equation:

$$r\dot{\phi}^2_{max} = \frac{4\ mgr^2}{I}, \tag{6}$$

and the maximum radial force exerted on the bar is obtained as:

$$F_{max} = mg(1 + \frac{4\ mr^2}{I}). \tag{7}$$

Since $I = I_r + mr^2$ the maximum force may be estimated as:

$$F_{max} \leqq 5\ mg.$$

Experiments carried out by Bauer [1976] with a strain-gage equipped high bar revealed that the maximum force exerted during a giant swing is $F_{max} \cong 4$ mg. It occurs when passing under the bar. Substituting this value into the left hand side of equation (7) we obtain:

$$I \cong \frac{4}{3} mr^2, \qquad r\ddot{\phi}^2_{max} \cong 3\ g. \tag{8}$$

Step 2: Calculation of the static joint moments at $\phi = 90°$. Assuming the model is kept at rest in the horizontal position ($\phi = 90°$) and all derivatives of equations (1) to (3) are set to zero, the equations for the joint moments simplify to:

$$\begin{aligned} M_\psi &= m_3 r_3 g \\ M_\theta &= m_2 r_2 g + m_3(r_3 + l_2)g \\ M_R &= m_1 r_1 g + m_2(r_2 + l_1)g + m_3(r_3 + l_1 + l_2)g, \end{aligned} \tag{9}$$

which can easily be verified in figure 9.

The first and second equation of (9) can also be used to calculate the magnitude of the hip and shoulder torques during the gymnast's static position of figure 10.

Step 3: Approximate calculation of dynamic hip joint moments. Assume that the mechanical gymnast is rotating with constant angular velocity $\dot{\phi}_{max}$

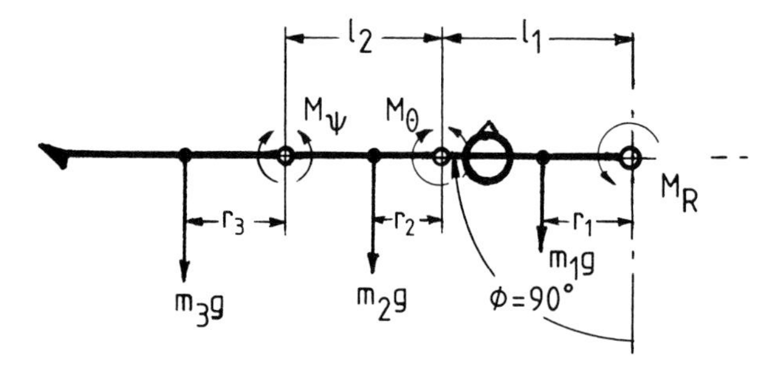

Fig. 9. Approximate calculation of static-joint moments at $\Phi = 90°$.

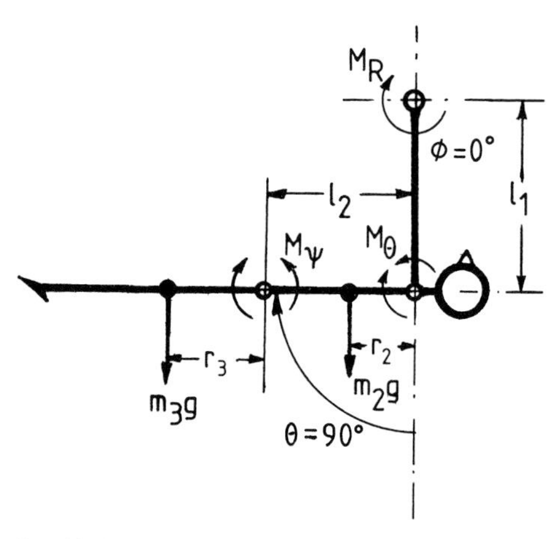

Fig. 10. Approximate calculation of static joint moments at $\Phi = 0°$ and $\Phi = 90°$.

calculated from equation (6) and that his spatial configuration remains constant according to figure 11 with a constant hip angle of $\psi = 90°$ during rotation.

Using equations (2) and (3) we see that the non-zero angular velocity brings additional terms in the moment equations which can be calculated as:

$$M_\psi \cong m_3 r_3 (g + (l_1 + l_2)\dot{\phi}^2_{max})$$
$$M_\theta \cong m_3 r_3 (g + l_1 \dot{\phi}^2_{max})\,. \qquad (10)$$

Step 4: Calculation of the moment of inertia with bent shoulder joints. Under the assumption that the configuration of figure 10 is constant during rotation or oscillation, it may be shown that the three equations of motion, (1) to (3), simplify to:

$$I^*\ddot{\phi} + mgr^*\sin(\phi + \alpha) = 0, \tag{11}$$

where

$$I^* = I_1 + I_2 + I_3 + m_1r_1^2 + m_2(r_2^2+l_1^2) + m_3((r_3+l_2)^2+l_1^2)$$

$$r^* = \frac{\sqrt{(m_1r_1 + (m_2+m_3)l_1)^2 + (m_2r_2+m_3(r_3+l_2))^2}}{m}$$

$$\tan\alpha = \frac{m_2r_2 + m_3(r_3+l_2)}{m_1r_1 + (m_2+m_3)l_1}. \tag{12}$$

Step 5: Approximate calculation of the shoulder joint moment after an assumed instantaneous changeover from straight body position to the bent body position of figure 10 at $\phi = 0°$. Consider the model is doing a downswing from the handstand position and going under the bar with angular velocity $\dot{\phi}_{max}$ (equation 6). At this point the assumed instantaneous changeover from straight hang to the position of figure 10 takes place. The angular velocity after changeover can be calculated from equation of conservation of angular momentum which gives:

$$\dot{\phi}^* \cong \dot{\phi}\frac{I}{I^*}. \tag{13}$$

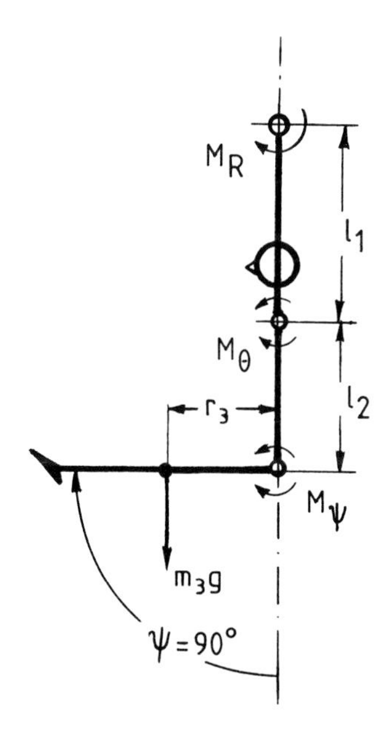

Fig. 11. Approximate calculation of dynamic hip joint moments.

Therefore, equations (2) and (3) gives:

$$M_\psi = m_3 r_3 (g + l_1 \dot{\phi}^{*2})$$
$$M_\theta = (m_2 r_2 + m_3 (r_3 + l_2))(g + l_1 \dot{\phi}^{*2}) \,. \tag{14}$$

Step 6: Approximate calculation of the radial forces after the assumed instantaneous changeover of step 5. From equation (8), we have $I = \frac{4}{3} mr^2$. Assuming we have for $I^* = \frac{4}{3} mr^{*2}$ and estimate $r = \frac{4}{3} r^*$, we obtain from equations (6) and (7):

$$r^* \dot{\phi}^{*2} = \frac{64}{9} g \cong 7\, g,$$

and

$$F^*_{max} \cong 8\, mg. \tag{15}$$

The value of eight times the body weight does not occur in reality for a real gymnast. However, since we want to calculate the possible reaction forces when the model is controlled by hand, we must take this case into consideration.

Derivation of the Equations for Experiments

The design of a physical model is an interaction between mathematical calculations and experimental work. Therefore, equation (4) will be used to yield an experimental method in order to determine the coordinates of the center of gravity and the moment of inertia of an arbitrarily shaped physical pendulum [Schmidt, 1961].

At first let the pendulum oscillate around its horizontal axis and measure the time period of oscillation T_1 with a stop watch. After attaching a string of length a from the link to the horizontal axis and measuring the time period T_2 (fig. 12a, b) the angular frequency of a pendulum described by equation (4), is determined by:

$$\omega_1^2 = \frac{mgr}{I_r + mr^2}\,; \quad \omega_2^2 = \frac{mg(r + a)}{I_r + m(r+a)^2}\,, \tag{16}$$

and the time period of oscillation is:

$$T_1 = \frac{2\pi}{\omega_1}\,; \quad T_2 = \frac{2\pi}{\omega_2}\,. \tag{17}$$

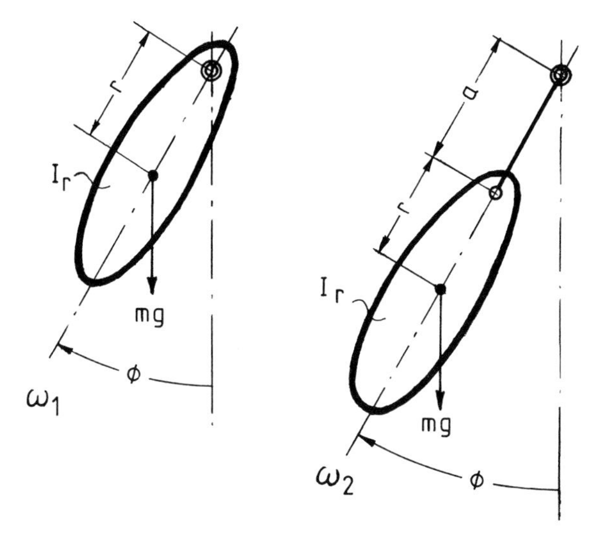

Fig. 12. Experimental setup for parameter identification.

Substituting ω_1, ω_2 into equation (17) and eliminating I_r we obtain the distance of the center of gravity as

$$r = \frac{a(4\pi^2 a - gT_2^2)}{g(T_2^2 - T_1^2) - 8\pi^2 a} . \tag{18}$$

To find the local term I_r of the moment of inertia, ω_1^2 must be substituted into equation (17), which after rearranging, yields:

$$I_r = \frac{mgrT_1^2}{4\pi^2} - mr^2 . \tag{19}$$

Since each link of the three link model may be suspended as a one-link pendulum their moments of inertia and center of mass can be found by two time measurements according to figure 12a, b and equations (18) and (19).

Development of a Prototype

Fundamental Design Values

A dynamic model which is to be manually controlled must have control parameters compatible with the capacity of the human perceptual-motor system. High frequency model dynamics demand fast control actions and large models require high actuating forces. When a model is controlled by hand, the reaction time mechanism of the human operator determines the limits of the frequency range within which he can interact while his muscular strength sets the limits of the manipulation forces he can generate.

If the model is of small size, it is of low weight and thus easily transportable. However, the angular frequency of its swing motions will then be high. Any increase in size reduces its angular frequency and makes it more controllable, however requiring greater actuating forces. So some compromise must be reached to overcome the inconsistency between transportability and controllability. To gain some insight into the nature of this compromise, previously established equations will be used for calculating fundamental design parameters.

Equation (8) gives $I = \frac{4}{3}mr^2$. Since the expression $I_r + mr^2$ of equation (16) may be substituted by I, the angular frequency to be expected is obtained as:

$$\omega_l = \frac{1}{2}\sqrt{\frac{3g}{r}}. \tag{20}$$

The effect of an increase in spatial dimension by a factor of 4 can be calculated by substituting r of the above equation by 4r. The result obtained is:

$$\omega_l{}^* = \frac{1}{4}\sqrt{\frac{3g}{r}},$$

which represents a reduction of angular frequency by one-half of the original value. However, this will also mean that there is an increase in volume by a factor of 4^3. Assuming the same material is used for both models, the mass of the larger model will increase by the same factor with the result that there will considerable increases also in radial forces (equations (7) and (15)) and joint torques (equations (9), (10) and (14)).

In mechanical engineering a controlling hand force of 100 N is considered a reasonable value. Thus, equation (15) gives the maximum mass for the mechanical gymnast when setting $F^*_{max} = 100$ N as:

$$m \cong 1.25 \text{ kg}.$$

The center of mass of a gymnast in straight body position with straight arms over his head is located approximately 40% of his total length away from his hands. Selecting a stretch length[1] for the model of 70 cm, then the distance r from the point of support to the center of mass can be determined as:

$$r \cong 28 \text{ cm}.$$

Substituting that value into equation (20) with $\omega_1 = \frac{2\pi}{T_1}$ the time period of oscillation can be calculated as:

$$T_1 = 4\pi \sqrt{\frac{r}{3g}} \cong 1.2 \text{ s},$$

which is approximately six times slower than the average reaction time of the human being and therefore lies within the frequency range of a human operator for control activities.

The model to build is a reduced size copy of a real gymnast of 75 kg mass and 175 cm height, whose total length is about 220 cm with arms stretched over his head. Compared to the total model length of 70 cm we have a reduction factor of:

$$k = \frac{220}{70} \cong 3.2.$$

The average density of a human is $\rho \cong 1.1 \frac{10^3\text{kg}}{\text{m}^3}$ this gives the volume of the real gymnast as:

$$V_{rg} = \frac{75 \text{ kg m}^3}{1.1 \ 10^3\text{kg}} = 0.068 \text{ m}^3 .$$

The volume of the model can therefore be calculated as:

$$V_m = \frac{V_{rg}}{k^3} \cong 0.002282 \text{ m}^3 ,$$

and the density of the construction material to select is:

$$\rho = \frac{m}{V_m} \cong 0.55 \frac{10^3\text{kg}}{\text{m}^3} ,$$

which is a value within the density range of wood.

[1] The selection of the stretch length is based on intuition in combination with a trial and error process during design, depending on the dimensions of commercially available construction material.

Construction Details

Model Description. Figure 13 shows the front view of the mechanical gymnast on a highbar which consists of a bottom plate (28) with two cylindrical supporting posts (3, 11) and a metal tube (4) as bar. The metal tube is pivoted at both ends in the supporting posts. Each of the two bearings consists of a sack-hole which can take axial and radial forces.

The model gymnast is hanging on the bar. Both cylindrical arms (6, 10) are fixed to the bar by two screws (5, 9) and to the shaft of the shoulder joint by two nails (51, 74). The trunk (53), with the head, is pivoted on the shaft (73) of the shoulder joint. Both cylindrical legs (63, 85) are fixed by two nails to the hip joint shaft (79) which is pivoted in the trunk.

Prestretched rubberbands (46, 47, 58, 59, 69, 70, 82, 83) fixed with metal rings to eye bolts (50, 52, 55, 57, 75, 76, 78, 81, 94, 95), cause the model to stay in a straight position when at rest. They also prevent the joints from over extension.

The string (42) responsible for the hip joint flexion consists of twisted perlon fibres. It is fastened with a swivel (60) to eye bolt (62) on the right leg (63) which is fixed to the left leg by the hip-joint shaft. Two nails (54, 77) prevent the legs from slipping out of alignment with each other.

Coming from eye bolt (62), string (42) is led behind the shoulder joint shaft, and then upwards in the plane of the arms into a hole bored in the horizontal bar. From there the string passes round a diversion pin (7) to run within the tubular bar and out through an outlet hole drilled through the supporting post (3). After going around pulley-block (43), string (42) is fixed at lever (35) with swivel (41) which keeps the string from getting twisted when the model gymnast is doing complete rotations.

Shoulder-joint flexion can be done with perlon string (14). It is fixed to a swivel (91) which is fastened to the bottom end of the trunk. String (14) is led from the trunk mounting through a hole drilled centrally in the horizontal bar, round the diversion pin (8) and along the inside of the hollow bar eventually to emerge at the outlet hole drilled in line with the bar centre, through the left supporting post. String (14) is then connected to swivel (15) which is fixed to lever (21). To reach best performance the pivot bearings of bar, hip and shoulder shafts, the swivel bearings and the string diversion devices must be designed to have low friction.

Figure 14 shows the arrangement of the operating levers (35, 21). The levers are pivoted with hingejoints (34, 22). The mounting points of string (42) and (14) are chosen in such a way that the maximum possible travel of the levers cause the maximum possible amount of joint flexion. To prevent

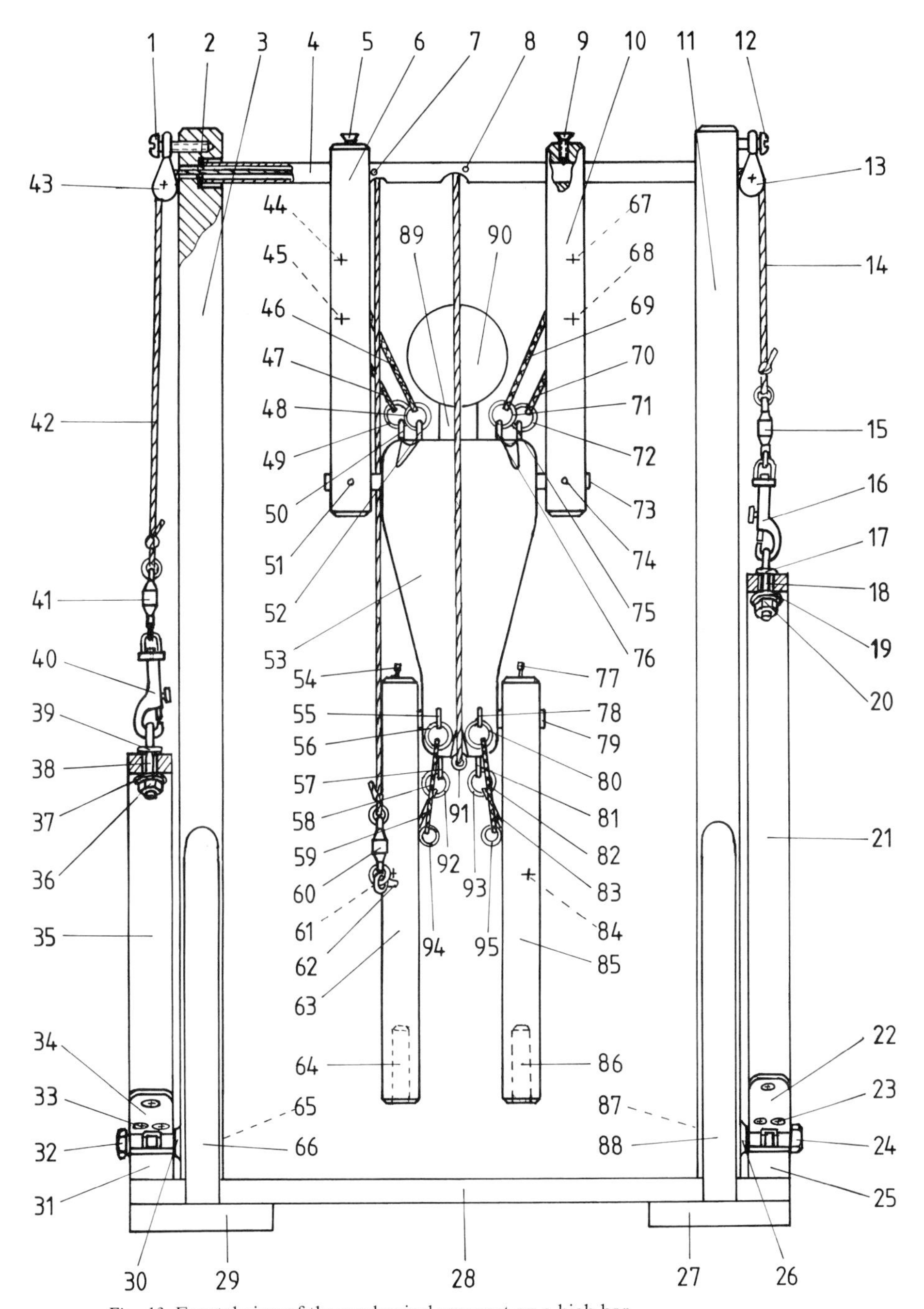

Fig. 13. Frontal view of the mechanical gymnast on a high bar.

Table I. List of parts (numbers marked with an asterisk * indicate parts which are hidden or do not lie in the drawing plane)

Specification	Material	Dimension	Quantity	Part. No.
Arm	broom-stick wood	255 × 28∅	2	6,10
Bar	brass tube	365 × 12∅ × 1	1	4
Base plate	wood fiber board	460 × 200 × 16	1	28
Base plate support	wood fiber board	800 × 100 × 16	2	27, 29
Counterweight	steel (thread rod)	45 × M10	2	64, 86
Diversion pin	copper nail	30 × 2	2	7, 8
Dowel	wood	30 × 8∅	2	97, 98
Dowel	wood	30 × 6∅	2	101, 102
Eye bolt (wood-thread)	steel	15 × 3∅, 11∅	10	50, 52, 55, 57, 75, 76, 78, 81, 94, 95
Eye bolt (wood-thread)	steel	25 × 3.5, 17∅	1	62
Eye bolt with nut	steel	23 × M5, 17∅	2	17, 18, 19, 20, 36, 37, 38, 39
Hat nut (metric thread)	steel	M4	2	111
Head	wood	70∅	1	90
Hinge	steel	65 × 30	2	22, 34
Hip-axle	brass tube	176 × 10∅ × 1	1	73
Key ring	steel	16∅	4	96, 96*
Leg	broom-stick wood	290 × 28∅	2	63, 85
Lever base	wood	60 × 30 × 20	2	25, 31
Metal ring	steel	18∅ × 2.5∅	8	48, 49, 56, 71, 72, 80, 92, 93
Nail	steel	30 × 1.7∅	2	51, 74
Nail	steel	50 × 2.3∅	2	54, 77
Neck	broom-stick wood	25 × 28∅	1	89
Operating lever	wood	1,300 × 30 × 20	2	21, 35
Pulley-block	brass	37 × 13 × 11, eye 6∅	2	13, 43
Reinforcing brace	wood	20∅, see fig. 18	4	65, 66, 87, 88
Rubberband	rubber	50∅ × 4 × 1	10	47, 58, 59, 70, 82, 83
Rubberband	rubber	50∅ × 1 × 1	10	46, 69
Shoulder-axle	brass tube	176 × 10∅ × 1	1	73
Spring safety hook	brass (galvanized)	75 long	2	16, 40
Spacer	aluminium tube	15 × 6∅ × 1,45°	6	109, 110, 109*. 110*, 113, 113*
String	twisted perlon fibers	800 × 3∅	1	14
String	twisted perlon fibers	800 × 2∅	1	42
Supporting post	broom-stick wood	750 × 28∅	2	3, 11
Swivel	steel	7∅	4	15, 41, 60, 91

Table I. (cont.)

Specification	Material	Dimension	Quantity	Part. No.
Thread rod	steel	105 × M4	2	105, 106
Trunk	wood	see fig. 15	1	53
Washer	brass	12∅ × 3	2	2, 2*
Washer	steel	8∅ × 2	1	91*
Wing nut	steel	M5	2	26, 30
Wing nut	steel	M4	6	107, 107*, 108, 108*, 112, 112*
Screws				
Flat-headed, countersunk, wood-thread	steel	20 × 4∅	2	5, 9
Flat-headed, countersunk, wood-thread	steel	17 × 4∅	6	23, 33
Flat-headed, countersunk, wood-thread	steel	30 × 4∅	6	23*, 33*
Flat-headed, countersunk, wood-thread	steel	25 × 4∅	8	99, 100
Round-headed, wood-thread	brass	35 × 5∅	2	1, 12
Round-headed, wood-thread	steel	20 × 3.5∅	7	44, 45, 61, 62, 68, 84, 91*
Hexagonal-headed, metric thread	steel	40 × M5	2	24, 32
Cylindrical-headed, metric thread	steel	50 × M4	4	103, 104, 103*, 104*

the model from tipping over when operated, the model must be firmly clamped down at front and back to its foundation (table, etc.). Joint flexion is achieved by pushing the operating levers downwards, extension is achieved by lifting them upwards. The upward lift releases the string and the recoil of the prestretched rubberbands results in extension back to the initial position.

The different parts are listed in table I. Numbers marked with an asterisk (*) indicate parts which are hidden or do not lie in the drawing plane. Numbers from table I which cannot be found in figure 13 should be traced in

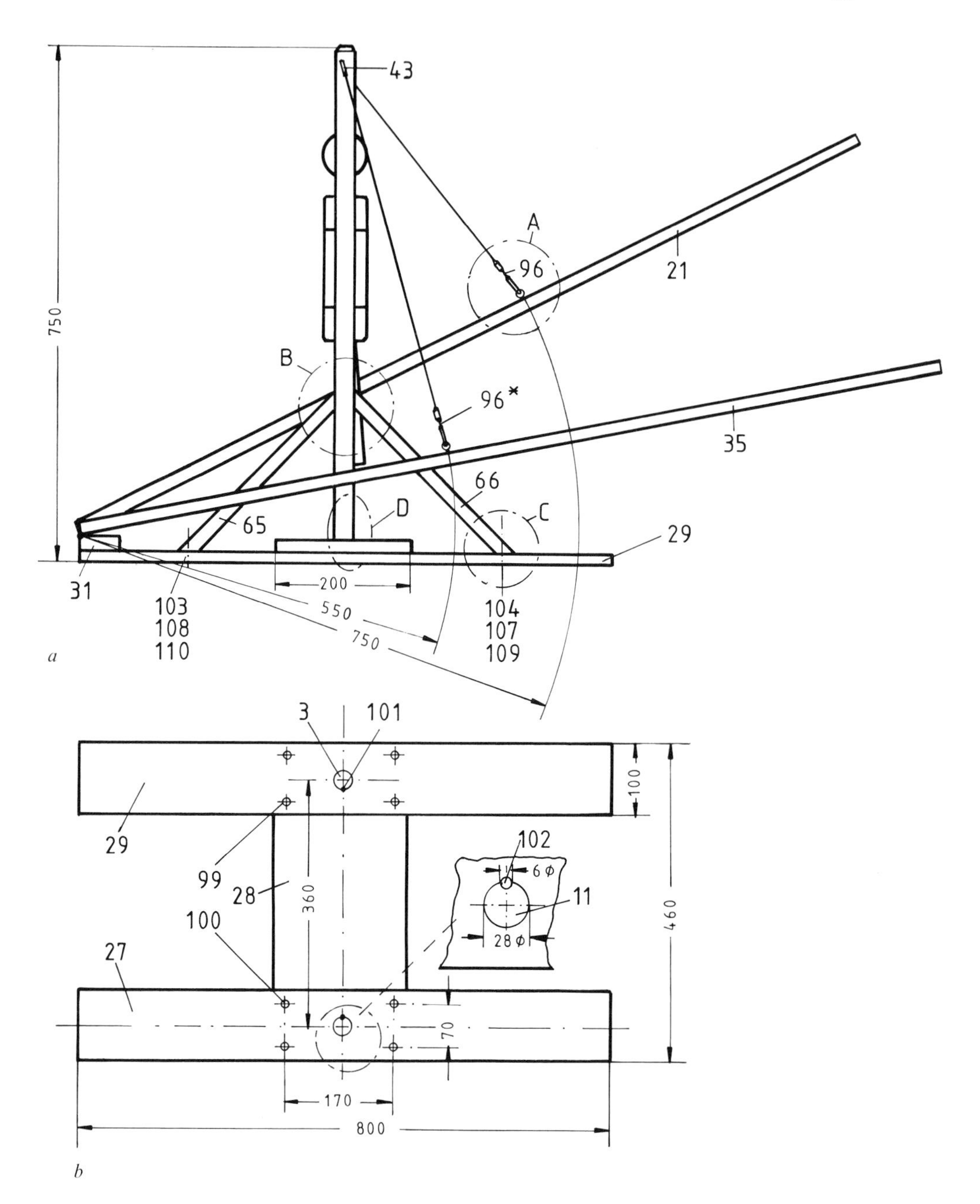

Fig. 14. Side and bottom views of the mechanical gymnast.

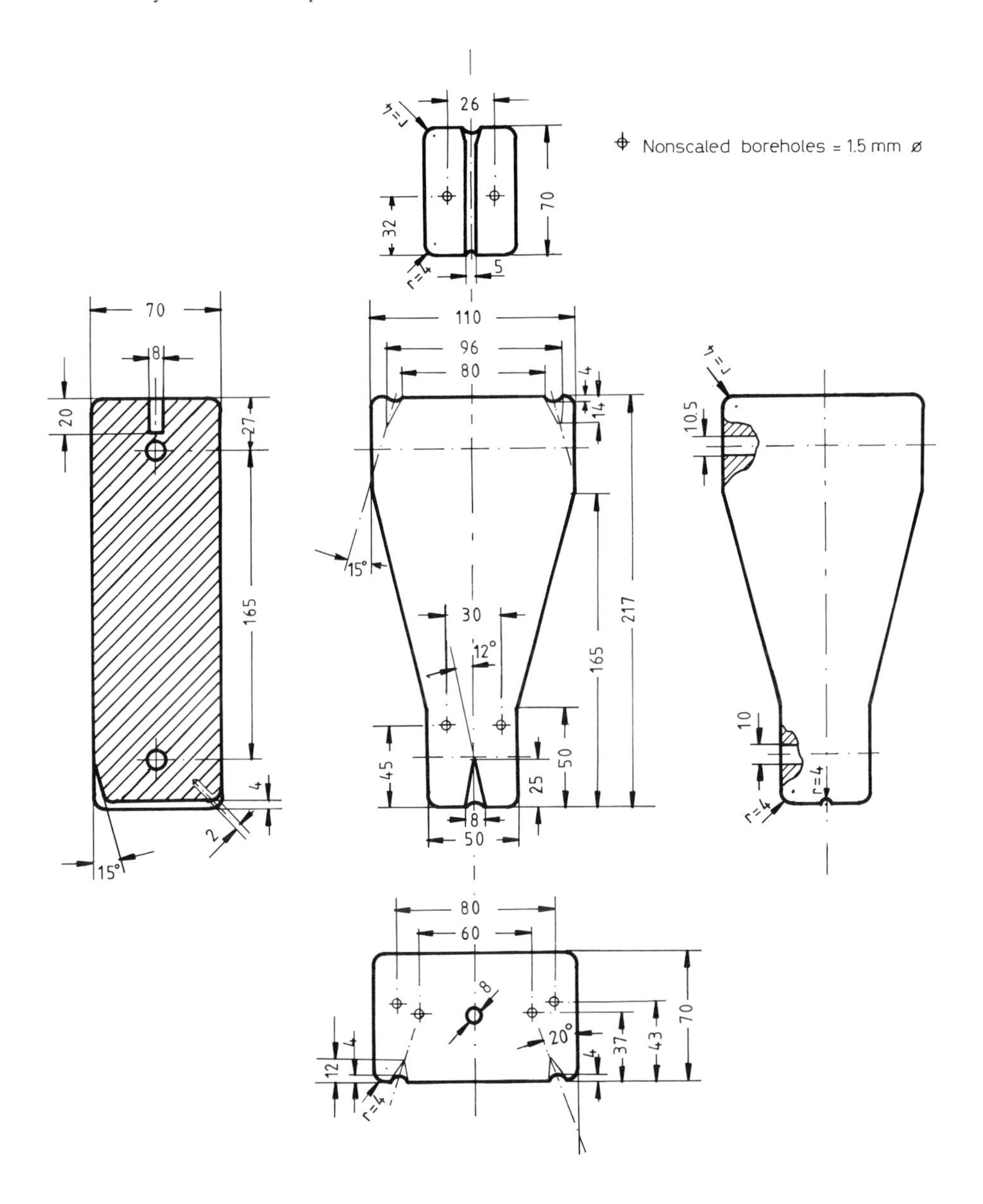

Fig. 15. Trunk dimensions.

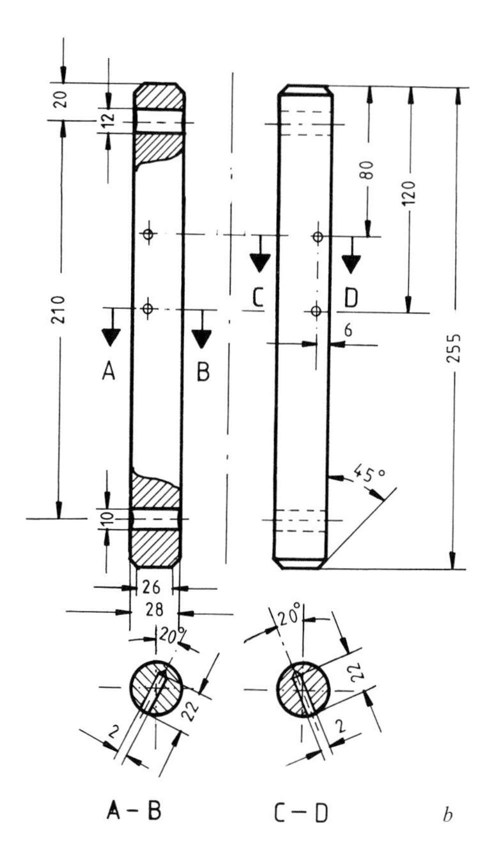

Fig. 16. Arm dimensions (dorsal view). *a* Left arm. *b* Right arm.

the figures 14 to 22. Washer (2*) is situated at the hidden end of the bar opposed to washer (2) (fig. 13). Number 91* marks a washer and a round headed screw which are used to fix one end of swivel (91) to the trunk. Screws (23*, 33*) are situated opposite to screws (23, 33).

Assembling Instructions. Perlon strings (14, 42) are attached to swivels (15, 91, 41, 60) with Palsteks. Each of the rubberband connections (47, 59, 70, 83) consists of two rubberbands (50 mmø × 4 mm × 1 mm) which are mounted in the following way as demonstrated on connection (47).

Both rubberbands are hooked to screw (45) then led through metal ring (49) and hooked to screw (45) again, so doubling the effective rubberband

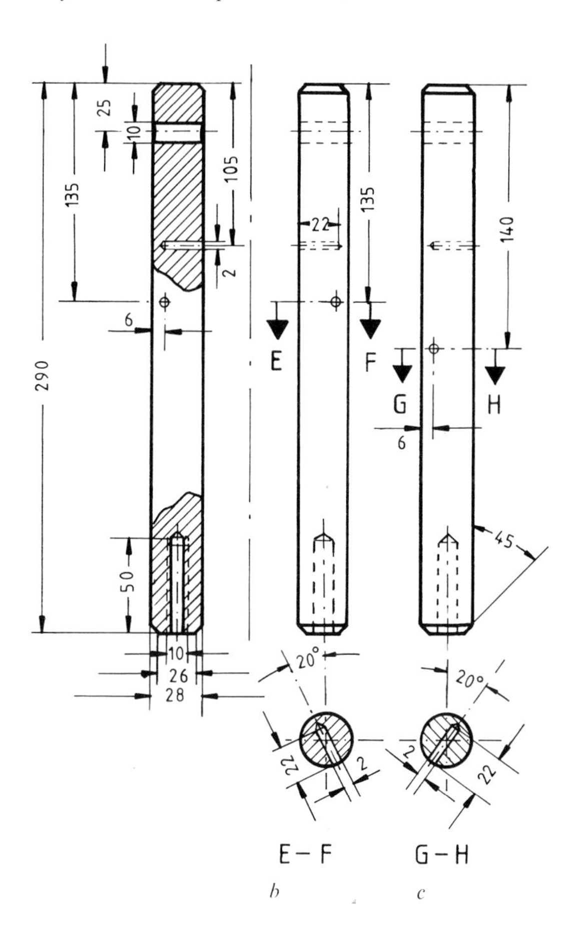

Fig. 17. Leg dimensions. *a* Left leg, dorsal view. *b* Right leg, dorsal view. *c* Right leg, frontal view.

forces. Each connection (46, 69) consists of five rubberbands (50 mmø × 1 mm × 1 mm) which are mounted differently in the following manner.

A ring of five rubberbands (46) is inserted into metal ring (48) for about 2 cm. The remaining long part of the rubberbands is then led through the eye formed by the two ends, then hooked to screw (44). No doubling of rubberband forces occur. Rubberbands (69) are mounted accordingly on screw (67). Connection (58) and (82) consists of one rubberband (50 mmø × 4 mm × 1 mm) each. They are mounted in the same way on metal ring (56) and (80) and then hooked into eye bolts (94, 95) whose eyes were opened with a screw driver.

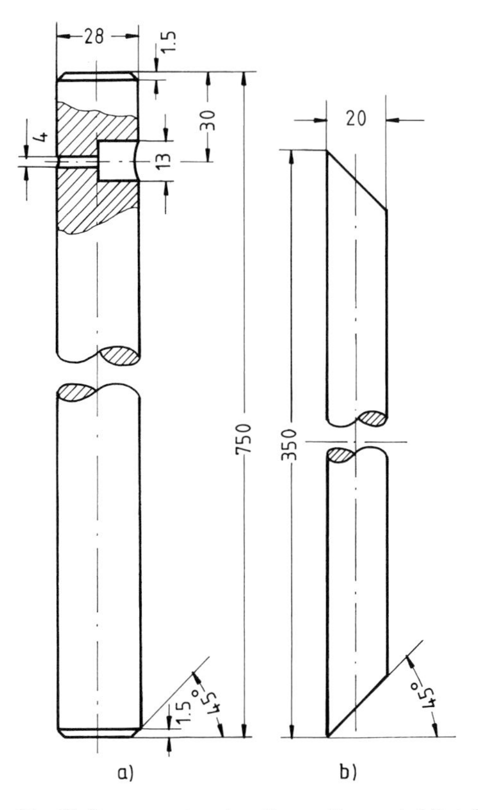

Fig. 18. Bar support parts. *a* Supporting post. *b* Reinforcing brace.

The most crucial part of the construction is the assembling of the diversion pins (copper nails 7, 8). Before inserting them into the respective housings in the bar (fig. 19, 22), each nail must be bent in the middle for about 90° without damaging its smooth surface. This is necessary to make sure that strings (14, 42) do not touch the rims of the bored holes when leaving the bar vertically. Strings should be lubricated from time to time to ensure minimal abrasion at the point of contact.

The axis of the drilled sockets for the supporting posts should have a slight inclination inwards (fig. 20D) to guarantee slight axial pressure on the bar from the bearings.

Model Parameters. The parameters of this model are listed below. The values m_i and l_i ($i = 1,2,3$) were measured using a weighing scale and a

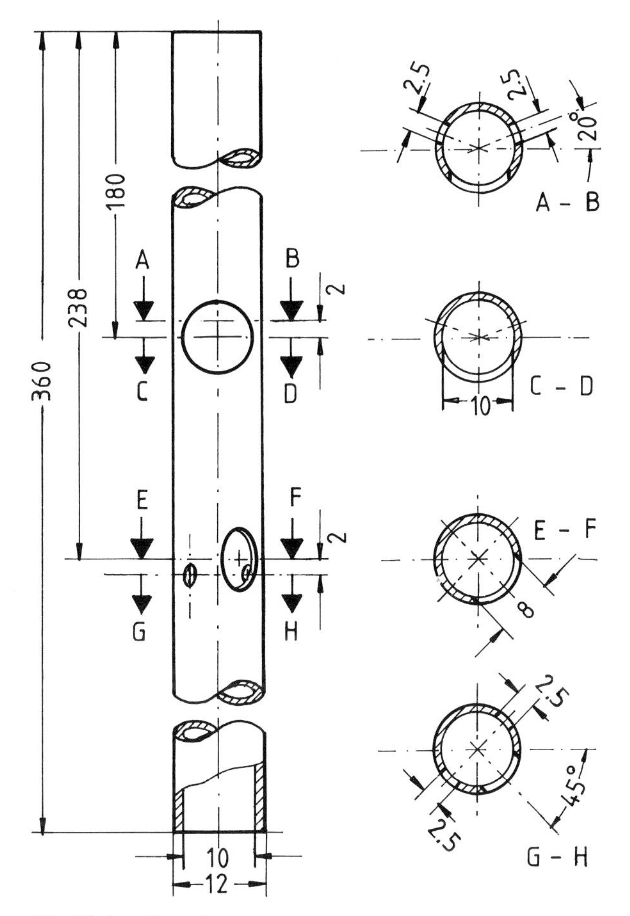

Fig. 19. Bar dimensions.

millimeter ruler, respectively. Each of the three links of the model – both arms plus the bar and the shoulder shaft, the trunk and the head, the legs with the hip shaft – was pivoted about its upper axis of rotation (fig. 12) and the two periods of oscillation (T_1 and T_2) were measured with a stop watch for small oscillations about the position of equilibrium. The distance r_i (i = 1, 2, 3) of the center of mass and the moment of inertia I_i were calculated with equations (18) and (19), respectively. By balancing each link on an edge the distance r_i was verified by a different method.

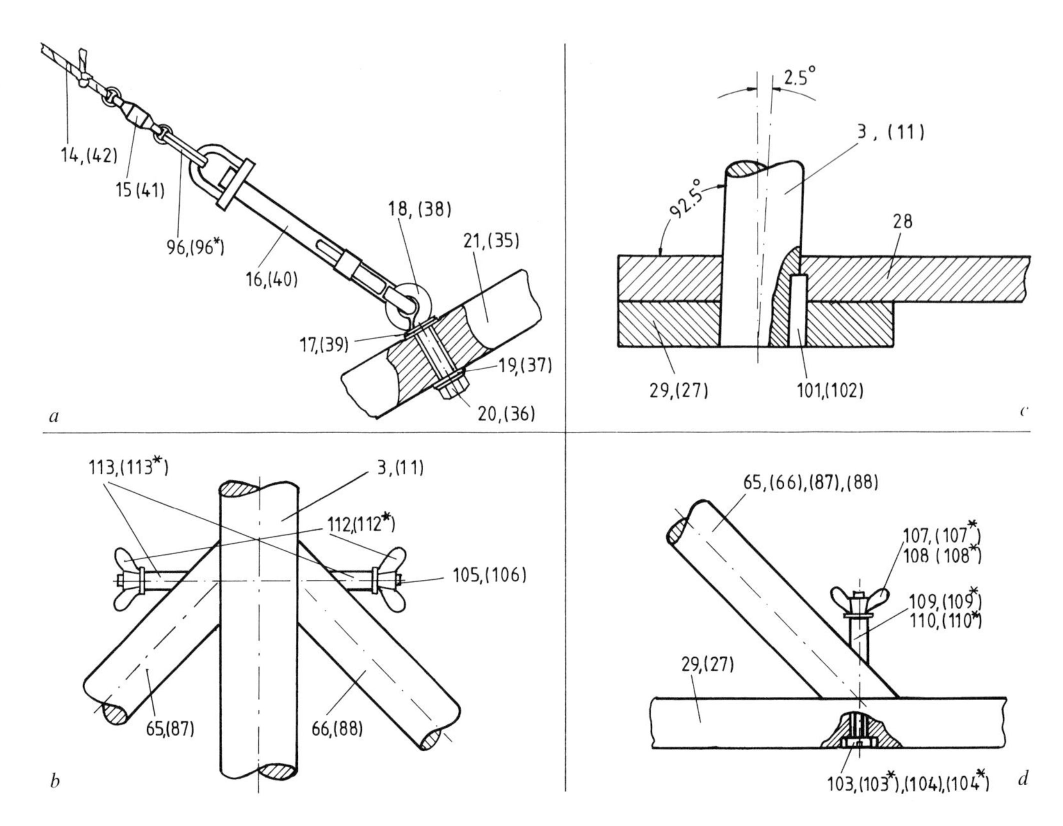

Fig. 20. Sector enlargements of figure 14. Sector a = Arrangement of spring safety hook and swivel on operating lever; sector b = installation of reinforcing braces on supporting post; sector c = installation of reinforcing brace on base plate support board; sector d = mounting of supporting post on base plate.

The model's parameters are:

$m_1 = 0.349$	$m_2 = 0.612$	$m_3 = 0.312$	(kg),
$r_1 = 0.071$	$r_2 = 0.05$	$r_3 = 0.13$	(m),
$l_1 = 0.212$	$l_2 = 0.165$	$l_3 = 0.265$	(m),
$T_1 = 0.78$	$T_2 = 0.9$	$T_3 = 0.87$	(s),
$I_1 = 1.987$	$I_2 = 4.629$	$I_3 = 2.424$	(10^{-3} kgm^2).

In order to verify these results the oscillation time and the moment of inertia relative to the bar for the assembled model was computed twice, once with the above parameters and once using two time measurements to permit the application of equations (18) and (19). The results obtained were as follows:

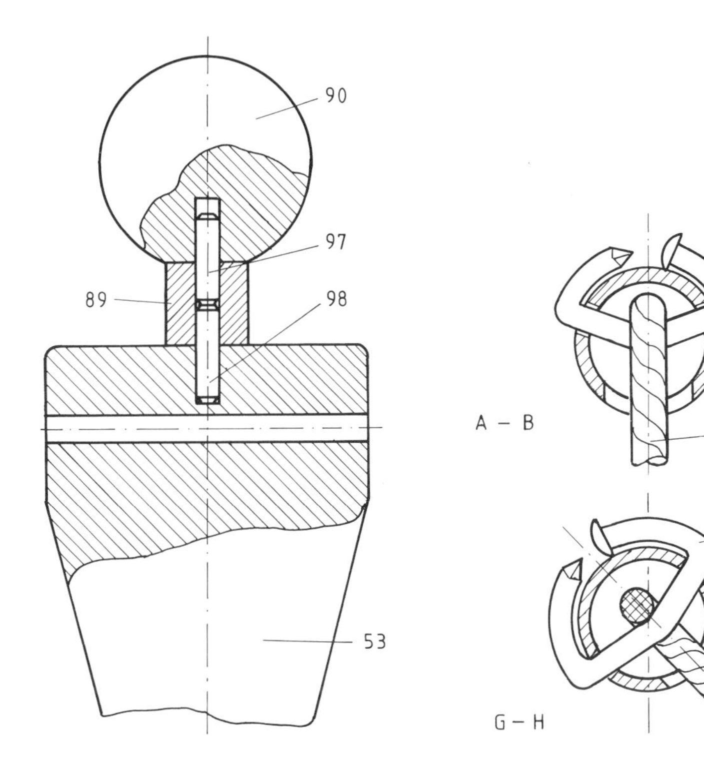

Fig. 21. Mounting of head, neck and trunk.

Fig. 22. String diversion arrangements.

$m = \Sigma m_i = 1.273$ kg $\quad i = 1, 2, 3$
$r = 0.27$ m
$T = 1.24$ s
$I = 0.131$ kgm^2 (moment of inertia relative to the bar).

In comparison with the human body, link 1 has very large values for mass and moment of inertia relative to links 2 and 3. This is due to the bar which must be fixed to the arms for mechanical reasons thus contributing to the mass and inertia of link 1 considerably.

Having identified the above listed parameters the left hand sides of equations (4) to (15) can be calculated in order to find the mechanical properties of the moving and nonmoving parts of the mechanical gymnast.

Figures 23–26 give an impression of how the model looks and how the rubberbands function as joint torque producers.

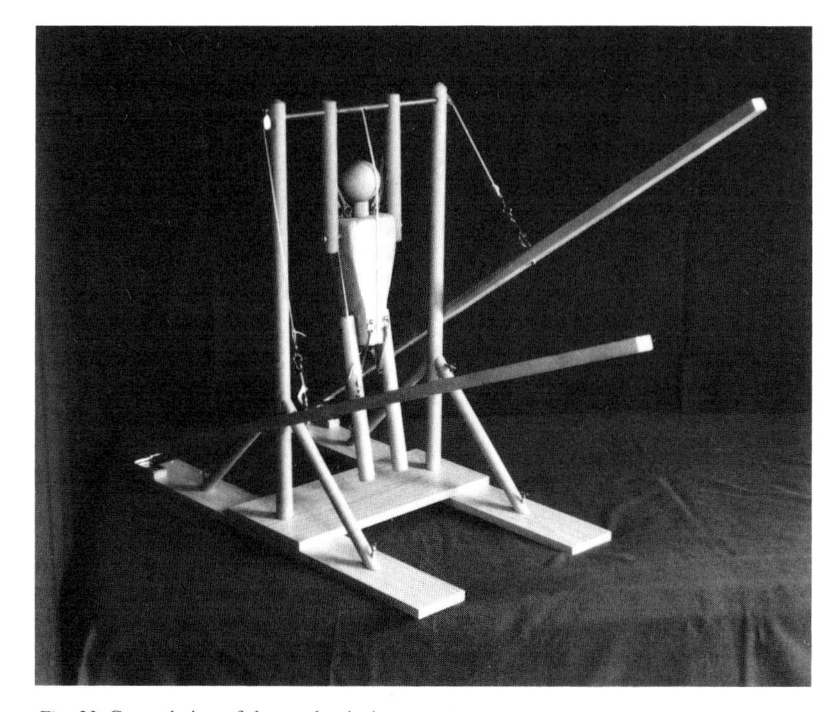

Fig. 23. General view of the mechanical gymnast.

Conclusion

The models presented here are moving mechanisms made of solid structures which have actually been built and used. The Bauer model has also been demonstrated to the International Society of Biomechanics. They belong to the category of physical or mechanical models which are used in sport biomechanics as an aid to understanding and as means of providing tactile and visual information on various sports techniques. In each case the human body is modelled by plane or three-dimensional multilink systems which are suitable for demonstrating either spatial components only (static models) or spatial, temporal and kinetic components simultaneously (dynamic models).

Dynamic models may be subdivided into two types according to their manner of control. One type which operates in an open loop manner after having set the initial conditions (fig. 6, 7a) and another type which requires

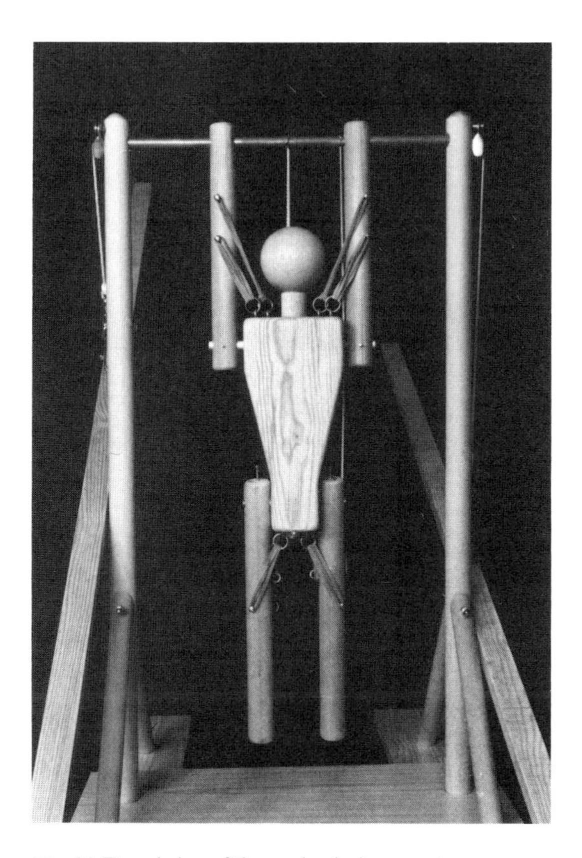

Fig. 24. Dorsal view of the mechanical gymnast.

closed loop interactions with a human operator to perform the motions designed for (fig. 5; 7b-d; 13).

Whilst the group of models first mentioned demonstrates desired single movements, the second set of models can be controlled interactively as an extension of the individuals perceptual-motor system and so initiate motor learning processes and promote internalisation of skills in a closed loop, error correcting manner similar to the learning process of real athletes. In particular this applies when working with the mechanical model illustrated in figure 13 from which a complex and familiar repertoire can be elicited.

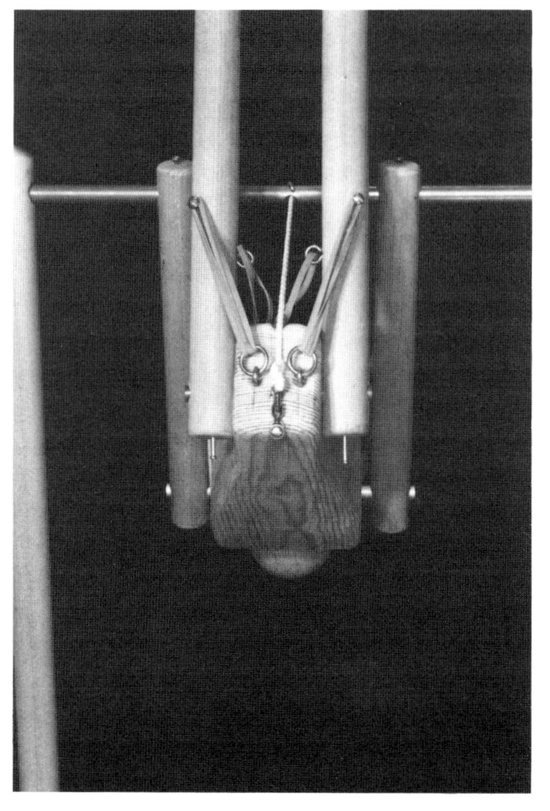

Fig. 25. Connection of rubber bands from trunk to legs.

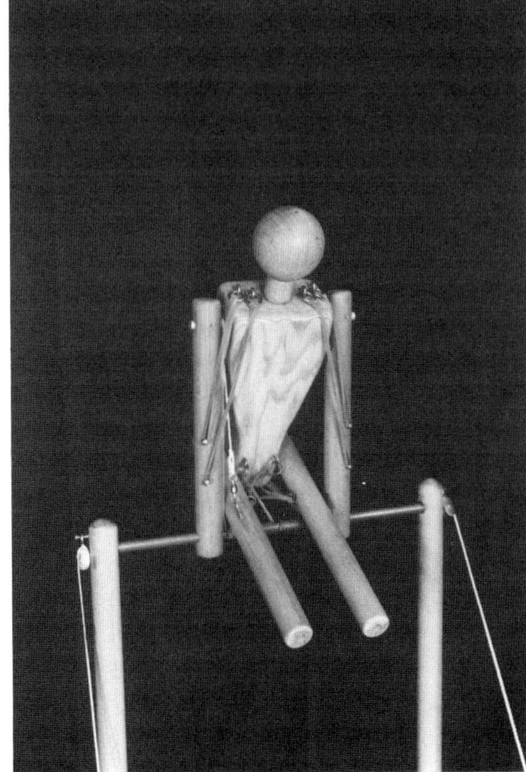

Fig. 26. Connection of rubberbands from trunk to arms.

References

Balster, K.: Der Einsatz eines Holzstrichmännchens im Sportunterricht. Sportunterricht *9:* 144 (1982).

Bauer, W. L.: Einsatz von Dehnungsmesstreifen zur Untersuchung von Lernvorgängen bei Turnübungen am Reck. Messtechn. Briefe *12:* 24–30 (1976).

Bauer, W. L.: Mathematical modelling and optimization and their influence on sport movements-possibilities and limitations; in Baumann, Internationales Symposium, Köln 1980. Biomechanik und sportliche Leistung; Schriftenreihe des Bundesinstituts für Sportwissenschaft, vol. 40, pp. 129–144 (Karl Hofmann Verlag, Schorndorf 1983).

Bauer, W. L.: A manually operated biomechanical model of an internally energized self-oscillatory system; in Winter, Norman, Wells, Hayes, Patla, Biomechanics IXB. International Series on Biomechanics, No. 5B, pp. 383–387 (Human Kinetics, Champaign 1985).

Boerchers, T.: Modellversuch – Sport an berufsbildenden Schulen. Abschlussbericht (Bremen 1986).
Cundy, H.M.; Rollett, A.P.: Mathematical models, 2nd ed. (Oxford University Press, Oxford 1961).
Fasol, K.H.; Jörgl, H.P.: Principles of model building and identification. Automatica *16:* 505–518 (1980).
Frohlich, C.: Die Physik der Schrauben und Saltos. Spektrum Wissenschaft *5:* 123–130 (1980).
Gercken, H.: Entwicklung und Anwendung eines sensomotorischen Lernprogramms für die Kenterrolle; Examensarbeit, Fachgebiet Sportwissenschaft (Universität Bremen, Bremen 1984).
Nickel, U.: Angewandte Bewegungslehre – Ausgewählte Phänomene sportlicher Bewegung im Experiment; Schriftenreihe des Bundesinstituts für Sportwissenschaft, vol. 48, pp. 64–65, 150 (Karl Hofmann Verlag, Schorndorf 1983).
Schmidt, R.: Mechanik einer Turnübung – Die Riesenfelge am Reck. Praxis Leibesübungen *12:* 139–144 (1961).

Dr.-Ing. W.L. Bauer, Sportturm, Zi. 3250, Universität Bremen,
Postfach, D-2800 Bremen 33 (FRG)

Med. Sport Sci., vol. 25, pp. 142–176 (Karger, Basel 1987)

Biomechanics in Gymnastics

G.-P. Brüggemann

Institut für Leichtathletik und Turnen
Deutsche Sporthochschule, Köln, BRD

Introduction

Gymnastics appears to be growing ever more popular. The persistent search by the gymnast for more advanced skills and the effect this is having on his originality, burgeoning artistry and virtuosity, accounts in no small measure for this increasing popularity. Perhaps the greater risk taking that is involved, is also a factor. Certainly, the growing number of serious accidents and injuries associated with sport reflect its hazardous nature.

Biomechanics has much to contribute to the continued progress of sport. Its use may sharpen the understanding of the techniques that exist, suggest new skills that might be tried, and lead to the achievement of more advanced performances. It may also help in identifying and controlling circumstances that lead to accidents.

Modern-day training is now so intense that to increase the volume of gymnastic training any further would be to generate few if any benefits. Only through changing the quality of training can real improvements be expected. Such an emphasis on quality demands deep unterstanding and insight. Despite this, few investigations, well designed to provide this understanding, have been undertaken in gymnastics. As Dainis [1981] points out, "... until very recently the only sources of information on vaulting mechanics have been non-quantitative coaching books and articles".

Technical skill is only one component of gymnastics that must be understood. Another equally important component is that of mechanical load. Each single body movement, and thus each component of any sporting technique, places mechanical loads on bones, muscles, tendons and ligaments. If the difficulty of an exercise is to be increased, additional rotations and somersaults have to be added. Extra energy is then required. This leads to the forces involved being pushed to ever higher levels, so increasing the load on the body. For instance, recent intensive research has focussed on the physical stress imposed during landing and on ways of cushioning these impacts. Varying the technique of a given movement, or changing the apparatus used, also modifies the imposed loads. In the following pages these questions of technique and load will be examined separately. Nevertheless, it should be appreciated that both are closely interwoven and cannot be separated properly.

Techniques in Gymnastics

Gymnastic performance is composed of a variety of incompletely defined components of motion that have been the subject of persistent efforts at isolation, identification and classification with a view to achieving economy of understanding and efficiency in coaching. Perhaps the best known of these attempts are those reported by Leirich and Rieling [1967, 1968]. They extracted eight basic groups of elements from a process which differentiated exercises according to their starting and ending positions and the nature of the relative movements of associated body segments. These eight groups are as follows: (1) hip circle movements (felge movements); (2) kipping movements; (3) uprise movements; (4) rolling movements; (5) overturn movements; (6) jumping movements (bouncing movements); (7) upward swing and circle movements, and (8) leg swing movements.

Categorising movements in the way described by Leirich and Rieling, generated in Germany widespread discussions about the language of gymnastics and led to significant modifications in gymnastic terminology. Elsewhere, however, little attention appears to have been paid to their work in international gymnastics.

A less academic but more pragmatic approach was taken by Thevenin [1979]. Aiming at clear but distinct designations of the gymnastic movements, Thevenin differentiated all movements into nine categories as follows: (1) swings in hang; (2) swings in support; (3) horizontal leg circles; (4) rolls and somersaults; (5) twists; (6) dislocations; (7) grip changes; (8) releases, and (9) position changes.

Many of Thevenin's categories correspond to ones found in the work of Carraso published in 1972. In spite of all these efforts, coaching books on gymnastics usually stick to the special demands of a particular apparatus and normally differentiate performances, if at all, into categories of jumping, hanging and supporting events.

Modern gymnastics has been the subject of intensive and creative thought, and movements previously characteristically performed on only one piece of apparatus have now been modified and transferred across a range of available apparatus. Biomechanical insights, therefore, now demand not only an explanation of the nature of the movement but also details about the characteristics of the performer and of the properties of the apparatus and surroundings. This view underlies the following discussion and has led to the decision to restrict the paper to a discussion only of those matters for which there are: (a) supporting biomechanical data; (b) properly derived principles which have influenced gymnastic coaching, and (c) attributable changes in modern gymnastic technique.

In consequence, discussion will be restricted to take-offs from the floor, to vaults and to performances on the rings, but will exclude aerial mechanics or performances on the horizontal or uneven bars.

The Take-Off for the Backward Somersault from the Floor

The take-off for the backward somersault is one of the most important floor techniques. The most difficult movements in floor exercise depend upon the efficient execution of this particular transition phase in performance. The backward take-off initiates the linear and rotational impulses for the tucked double back, the double back with straight body (lay-out), the double, triple and quadruple twist, as well as the so-called Tsukahara (double back with full twist) and the double back with double twist.

A review of the literature indicates that only a few take-off studies have been reported of analyses of gymnasts of a suitably high technical standard.

Payne and Barker [1976] and Feller [1975] analysed the kinematics and kinetics of the backward somersault take-off. More differentiated studies regarding body segment movement patterns during backward take-off on the floor are reported by Brüggemann [1983a, b, 1984], who focussed his attention on the nature of body segment movement patterns during take-offs in floor exercises with a view to determining the contributions of the arms, the trunk and the legs to total angular and linear momentum of the body. In particular, he sought to define the role of the arms in the rotation of the body and in the achievement of height during flight.

In the following pages an outline is given of the results of several studies by the author. These involved a total of more than 60 highly skilled gymnasts and were devoted to the analysis of floor take-offs recorded during both training sessions and competitions. In studies conducted during training, use was made of the Kistler force platform and of kinematic recording facilities, but in studies conducted during competition only cinematographic records could be made. In both groups of studies, recordings began from shortly before final ground contact in the flic-flac or round-off approach. The procedures used were consistent across all the studies discussed. Movements were filmed from a distance of at least 15 m using a Locam 16 mm camera at a nominal 200 Hz. The ground reaction force vector, its point of location and the moment about the vertical axis were recorded from a Kistler platform. Segment masses were calculated from the regression equations of Clauser et al. [1969] and segment moments of inertia were computed according to a modified Hanavan [1964] model. The digitized segment endpoint data were smoothed using a digital filter with a cutoff frequency of 8 Hz. Linear and angular velocities and accelerations were computed via finite difference differentiation.

The angular momentum of the body segments relative to the body's centre of mass about the transverse axis is:

$$H = I\omega + rm\dot{r},$$

where I is the segment moment of inertia relative to the transverse axis through the centre of gravity of the segment, ω is the segment angular velocity, r is the vector locating the segment centre of mass relative to that of the body's, $\dot{r}$ is the time derivative of r, and m is the mass of the segment. The angular momentum of the total body with respect to its centre of mass is computed by the sum of the angular momenta of the segments. The inertial segmental forces are calculated from the segment mass and its linear acceleration. For comparisons between subjects, the take-off time was normalized.

Table I presents some selected results. The pre-flight phase of the round-off or of the flic-flac is seemingly short, lasting less than 100 ms. The horizontal velocity of the gymnast's centre of mass at touch-down is approximately 4 m/s and decreases during take-off. The vertical touch-down velocity was calculated at –0.4 m/s. During rebound the vertical velocity of the centre of mass increases to reach a take-off velocity of 4.1 m/s and more (fig. 1). A large angular momentum of –128 kg m^2/s about the transverse axis through the centre of mass of the body was also found at touch-down. During rebound this angular momentum is reduced by about 50%. The take-off changes the direction of flight of the centre of mass by about 60° and drastically decreases the initial angular momentum.

Table I. Summary of force-time, angular and linear velocity data

	Mean (n = 40) somersault	Mean (n = 26) double somersault
Duration of pre-flight, s	0.108	0.094
Vertical velocity at TD*, m/s	–0.600	–0.420
Horizontal velocity at TD, m/s	3.750	4.120
Angular momentum at TD, kg m²/s	–113.000	–128.900
Duration of take-off, s	0.131	0.123
(1) Maximum vertical ground reaction force, N	4,372.000	4,510.000
Minimum vertical ground reaction force, N	2,839.000	3,090.000
(2) Maximum vertical ground reaction force, N	6,069.000	6,846.000
Maximum horizontal ground reaction force, N	815.000	867.000
Minimum horizontal ground reaction force, N	–1,9000.000	–2,100.000
Takeoff hight (CM), m	1.165	1.160
Hight of CMs flight, m	0.953	1.074
Total hight of CM, m	2.118	2.234
Vertical take-off velocity, m/s	4.300	4.570
Horizontal take-off velocity, m/s	2.690	2.850
Takeoff angular momentum, kg m²/s	–55.900	–64.500
Takeoff angular velocity, rad/s	4.130	4.850

* TD = Touchdown

This important reduction of angular momentum in a typical double somersault is shown in figure 2. Because the body is free of external torques during the airborne phase, the total angular momentum should be constant. However, when angular momentum is computed during the flight phase, variations of about ± 5% are found. These are presumably due to measurement errors, particularly associated with the flexed trunk during the pike phase and the choice of inappropriate body segment parameters. Hay et al. [1977] reported similar problems in calculating a constant airborne angular momentum from film data.

The changes of linear and rotational momenta at take-off are caused by vertical and horizontal forces. The force-time curves shown in figure 3 are different from those of take-offs found in other sports. The touch-down of the legs on the damping mat after the approach flic-flac, produces only modest peaks in both the horizontal and vertical direction. These are in

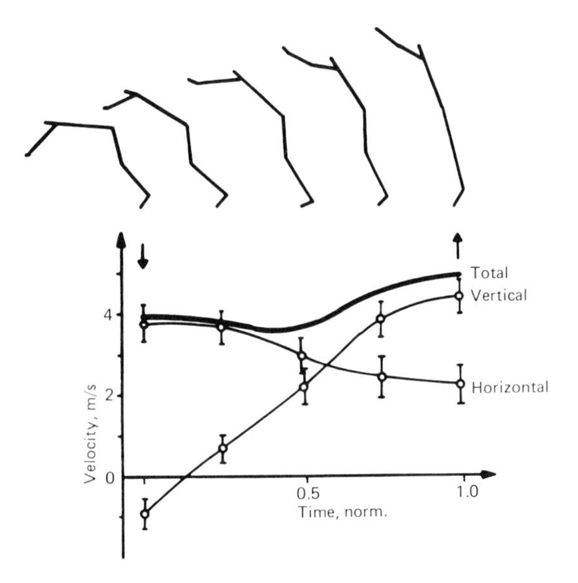

Fig. 1. Vertical, horizontal and resultant velocity of a gymnast's centre of mass during a backward somersault take-off (mean and SE of 26 subjects). The take-off time is normalized.

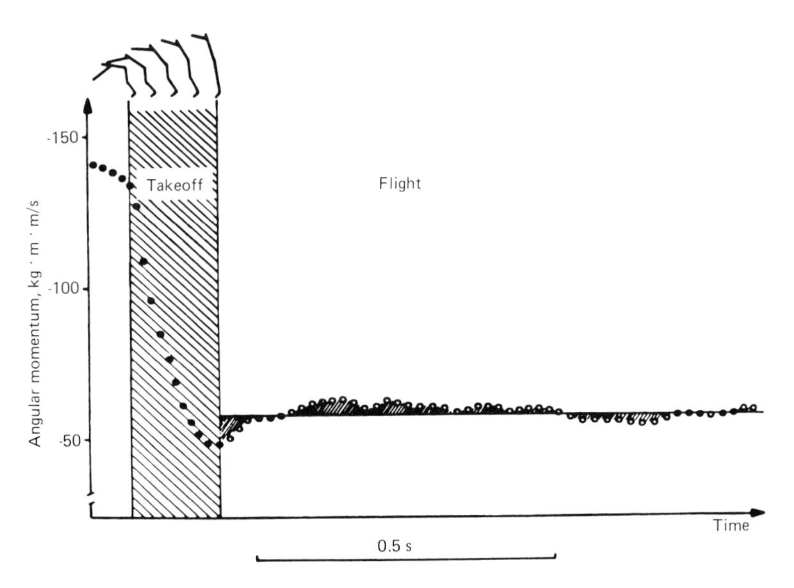

Fig. 2. Angular momentum about the transverse axis through the gymnast's centre of mass during pre-flight, take-off and flight in a typical double backward somersault.

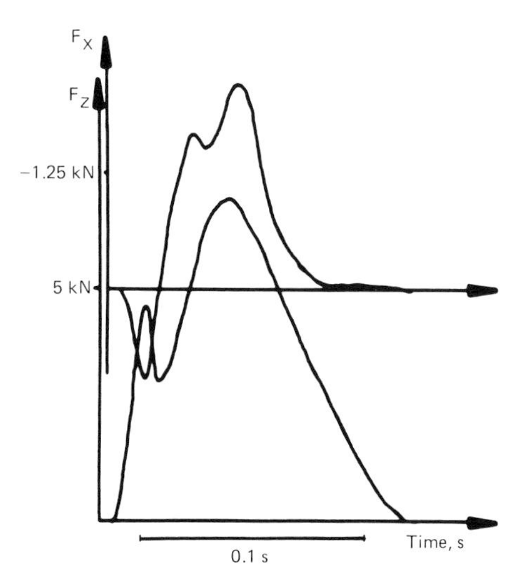

Fig. 3. Vertical (F_Z) and horizontal (F_X) ground reaction forces during a typical backward somersault take-off.

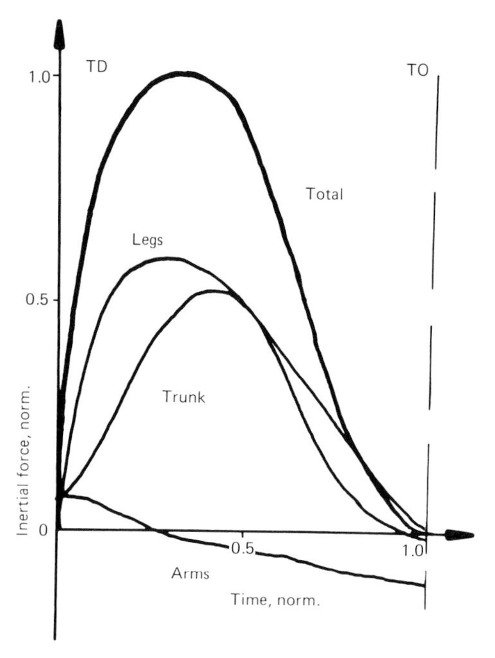

Fig. 4. Vertical inertial forces of the major body segments, normalized for body mass, during a backward somersault take-off (mean of 26 subjects). Take-off time is normalized.

marked contrast to the subsequent peaks. The double footed take-off drive which quickly follows, generates a high peak force of 6,800 N vertically and 2,100 N horizontally. It is these forces which generate the impulse which so markedly reduces angular momentum.

As vertical velocity and angular momentum in the somersault plane are to be optimized it is necessary to consider the contributions made to these parameters by the body segments. The vertical inertial force ($F_i = m_i a_i$) for the legs, the trunk and the arms are presented in figure 4 for the take-off period of the double backward somersault. The curves characterize the mean of the measured performances of 26 highly skilled gymnasts [Brüggemann, 1983a]. The legs and the trunk, because of the high acceleration and the trunk's large mass, are responsible for the majority of the total impulse over the take-off period. The highest values of the leg's inertial force are found during the eccentric phase of leg extension acitivity. The maximum vertical acceleration of the trunk occurs later and is followed by small negative intertial forces at the final phase of take-off.

The force-time curve associated with the arm movement is of major interest because this movement is often emphasized by coaches. Its force-time history is quite different from that of the other segments.

At touch-down the inertial force of the arms is positive and significant as, playing an apparently important role in the stretch-shortening cycle, they accelerate to load the leg extensors in the eccentric phase of their contraction. During the subsequent recoil and take-off drive, the arms reverse their role, producing a negative vertical inertial force which, combined with the low acceleration of the trunk, results in a reduction of the load on the shortening leg extensors. In the last take-off phase the arm action seems to influence the final extension of the knee and ankle joints.

A similar procedure was used to investigate the contributions of the segments to the total angular momentum of the body. Angular momentum about the body's principal transverse axis recorded from the same gymnasts showed a rapid fall largely parallel to the changes in linear momentum: the major change occurred in the first phase of the take-off following touch-down (fig. 5).

Angular momentum changes in both the trunk and legs have similar time histories with marked initial losses which level out at the end of take-off. In contrast, arm angular momentum remains approximately constant at first, even rising slightly, but then it decreases appreciably in the later stages of take-off. The final decrease seems to allow a differentiated modification of angular momentum of the whole body.

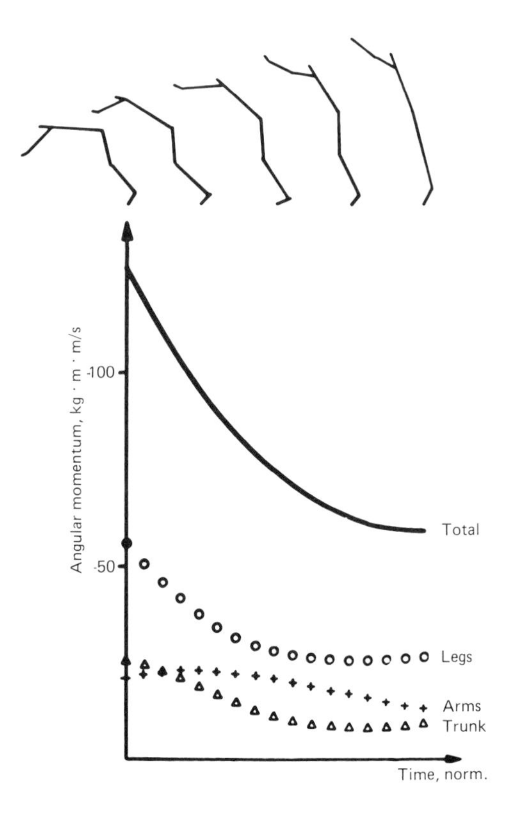

Fig. 5. Angular momentum contribution of the body segments to total angular momentum (solid line) with respect to the gymnast's centre of gravity during take-off for the double backward somersault. The angular momenta shown, include both local and remote contributions of the respective segments. The negative angular momentum is clockwise and consistent with the rotation direction of the backward somersault. The take-off time is normalized.

From the foregoing it is clear that the legs play the dominant role in contributing to total angular momentum during the first half of the take-off. The contribution of the legs, which is approximately twice that of the trunk, is almost entirely due to their large mass and their remote mass centre. Therefore, positioning the legs accurately at touch-down is of considerable importance in order to control the angular velocity of the body. The movements of the arms, so often stressed by coaches, are clearly much less important from a mechanical point of view, and the pattern of their contribution is quite different from that commonly described in standard coaching textbooks.

The relationships between the dynamic parameters measured at touch-down and the height achieved in the ensuing flight were determined in a group of 40 subjects using multiple regression analysis [Brüggemann, 1983a]. This analysis showed that angular momentum and horizontal velocity at touch-down were closely related to the height of flight during the somersault (r = 0.81).

In conclusion, a high linear velocity and a high angular momentum at touch-down are essential pre-requisites of a good back somersault, but primarily to ensure that the gymnast is poised in an optimal position to generate the impulse needed for the back somersault rather than directly to provide some of that impulse. Hence there is a need for great attention to be paid to the round-off or flic-flac approach.

The Rings

Techniques on the rings have developed rapidly since the 1960s when strength and static skills predominated and dynamic movements were generally performed poorly. During the 1970s and 1980s there was a marked change of vogue and dynamic swinging skills were developed until they became the hallmark of the highly skilled.

So fast has been this change that, although practical efforts were made to optimize dynamic techniques, no theoretical foundation for this optimization was laid. Indeed, the first quantitative study of ring work was that by Marhold [1961] on hip circle and backward uprise. However, even that study had equipment design rather than gymnastic theory as its goal. Later some minor studies into backward giant swing were reported by Dusenbury [1968], Fetz and Opavski [1968] and Peek [1968] together with some more substantial investigations into the kinetics of the felge-to-handstand by Sale and Judd [1974] and Nissinen [1983].

The following discussion extends the above work. It is based on work undertaken by Nissinen and Brüggemann [unpubl. report, 1982, 1986] at the European Championships of 1979, the German Championships 1980, and the European Championships for Juniors 1986, and centres both on the selection of techniques crucial to success in swinging on the stationary rings and also on their application to coaching.

In the above studies two Kistler force transducers were installed inside the metal framework on the rings just about the attachment of the ring cables. The force signals were registered on a PCM taperecorder and transmitted to a microcomputer for further analysis. The performances were filmed by a Locam camera with a nominal speed of 100 fps. The films were digitized using a 14-segment model of the human body. Segment masses, locations of centres of mass and segment moments of inertia were obtained by the regression equations of Clauser et al. [1969] and a modified Hanavan [1964] model, respectively. Backward and forward giant swings performed by gymnasts of various skill levels and ages, were investigated during training and laboratory sessions as well as during the above cited competitions. After digitizing, the raw data were smoothed by a digital filter prior to the calculation of segmental velocities and accelerations. For further analysis the gymnast's body was treated as a three-segment system consisting of arms, trunk and legs. Net muscle moments were calculated for the hip and the shoulder joints, respectively. According to Djatschkow [1974], every change of the basic swing movement of a multi-segment pendulum is directly related to the accelerating movements of the legs. Therefore, the relative movements of the legs with respect to the hips, of the trunk with respect to the shoulders, and of the arms with respect to the moving suspension point were investigated.

During the downward swing a first unweighting phase is to be indentified (fig. 6). There is a large forward displacement of the rings from the body centre of gravity to achieve a higher angular momentum. The achievement of

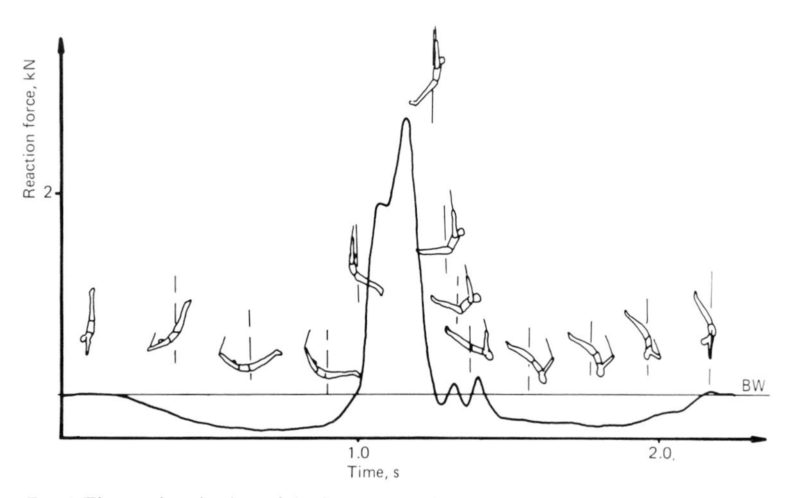

Fig. 6. The synchronization of the film and the force-time curve as measured by one of the two force transducers. The horizontal line represents half of the body weight (BW), the F-t curve represents a world-class gymnast during a straight-arm giant swing. The dotted lines indicate the vertical projection of the suspension point of the rings.

high angular momentum depends on keeping arms and rings parallel during the downward swing.

As the gymnast approaches the vertical position, the reaction force increases rapidly and surpasses that of body weight. Maximal peak reaction force is reached in the vertical position. The analysed trials have shown reaction forces of 6.5 to 9.2 times body weight. Those values are considerably higher than so far reported in the literature [Marhold, 1961; Sale and Judd, 1974]. This is due to changes in the technique of executing the downward swing.

After the peak, the reaction force decreases rapidly, returning to or below body weight. Technically superior trials show two smaller peaks above body weight in the reaction-force curve. These two peaks are explained by the effects of leg braking and trunk transfer, and then of trunk braking and arm initiation, respectively. These results contradict the coaching textbooks for teaching the giant swing. The often emphasized active push backward of the rings after the vertical body alignment, is not to be found in giant swings of high technical standard. A reaction force equal to that of body weight is not reached until returning once again to a handstand.

Details of the upward swing were further analysed by the relative segment movements and the resulting net muscle moments. Figure 7 illustrates the vertical velocities of the trunk, arm and leg segments respectively relative

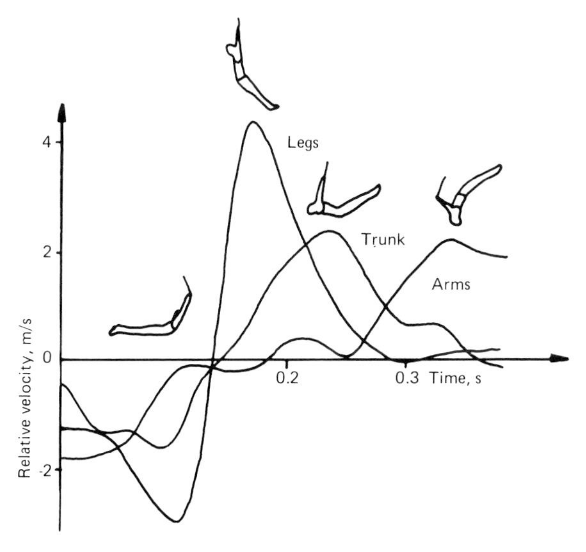

Fig. 7. The time history of the relative vertical velocities of the trunk, arm and leg segments during a straight-arm swing of a world-class gymnast.

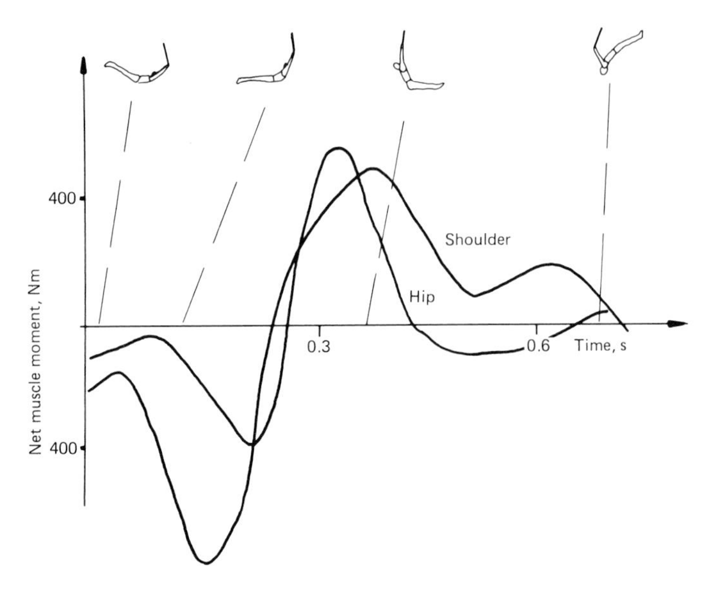

Fig. 8. The net muscle moments of the hip and shoulder joints during the most important part of the straight-arm giant swing of a world-class gymnast.

to the shoulder joint, to the rings and to the hip joints. After the vertical body alignment, figure 7 shows an increase of leg velocity to more than 4 m/s followed by a decrease to about zero. The braking motion of the legs leads to a complete cessation of movement of the segment relative to the hip joint. The corresponding curve of motion of the trunk increases later and less intensively. Those characteristics agree with the theoretical considerations of Djatschkow [1974]. The action of the arms occurs later and corresponds with the decrease of the trunk's relative velocity. The arm contribution begins after hip extension is completed. The coordination of flexion and extension of both shoulder and hip can be illustrated by the net muscle moments acting at the hip and shoulder joints (fig. 8). The curves indicate the pre-stretch phase of the shoulder joints. The flexion moment of the hip joint decreases, while the flexion of the shoulder joint increases. The extension of the shoulder joint occurs after the trunk has nearly reached a vertical position.

The above results again contradict the instructional books in gymnastics. The investigation has shown that the arms and shoulders should remain under the rings until hip extension in the upward swing has been fully completed.

Therefore, there is no active push of the rings forward or backward. From the foregoing it may be concluded that the initiation of the leg action (hip flexion-extension) is of primary importance for the coordination and execution of the swing. The active movement of the rings by the arms is relatively late in the swing after hip flexion-extension has been completed. This delayed arm action allows for a more favourable position of trunk and legs with respect to the point of support.

The results can be applied equally to backward or forward giant swings, simply by considering the relevant muscle moments as having opposite signs.

The Horse Vault

The discussion that follows is based on the work of Bajin [1979], Brüggemann [1979], Brüggemann and Nissinen [1981], Cheetham [1982], Dainis [1979], Dilman et al. [1985] and Nelson et al. [1985]. It will be presented in three parts. First, the experimental procedures common to all these studies will be identified. Second, an outline will be given of the classes of vaults. Third, the main principles of vaulting will be worked out using the most extensive set of data for the front-handspring so far published [Brüggemann and Nissinen, 1981].

All studies of vaulting used cinematography. The vaults were filmed by one, two or three high-speed cameras operating at 100–200 fps. All analyses were two-dimensional even if the camera set up allowed for three-dimensional calculations. All investigators smoothed the raw data with the same digital filter method using a similar cutoff frequency. Therefore, the results of the cited articles are comparable.

In high standard performances most vaults fall into the handspring, or Tsukahara group. These vaults include front-handspring with pre- or post-flight twists and may in addition also include post-flight somersaults. They also include the Cuervo vault. This vault consists of a half-twist followed by a one-and-a-half back somersault in post-flight.

The Tsukahara group involves performing a ½ twist during pre-flight followed by 1½ somersaults during post-flight with or without twist. In that group we classify the Kasamatsuvaults as well. This kind of vault shows ¼ twist during pre-flight followed by 1½ somersault with ¾ twist (or 1¾) in post-flight.

A round-off entry indicates that the gymnast performs a round-off onto the Reuther board. These vaults should be classified in the Tsukahara group.

But because of the special kinematic and kinetic characteristics of the take-off, the round-off entry vaults are grouped as an additional class of vaults. The essential vault begins with a backward body alignment relative to the approach direction. In the post-flight the gymnast completes a 1½ backward somersault with or without twist.

All vaults of these categories are subject to the same biomechanical principles.

The movement of vaulting can be subdivided into six phases: (a) approach; (b) take-off; (c) pre-flight; (d) support; (e) post-flight, and (f) landing.

The Approach

Run-up speed is measured as the horizontal velocity of the centre of mass of the body at initial contact with the Reuther board. This velocity increases with the level of performance, a trend clearly shown (fig. 9) by plotting the means and standard deviations of the approach velocities of 10 World, 10 National, and 10 Junior class German gymnasts [Brüggemann and Nissinen, 1981]. From the results one may conclude that the horizontal touch-down velocity is of high significance. The correlation coefficients between the run-up speed and the height and distance of the post-flight are calculated as $r = 0.78$ and $r = 0.83$, respectively. The analysis of variance shows high significant differences between all groups.

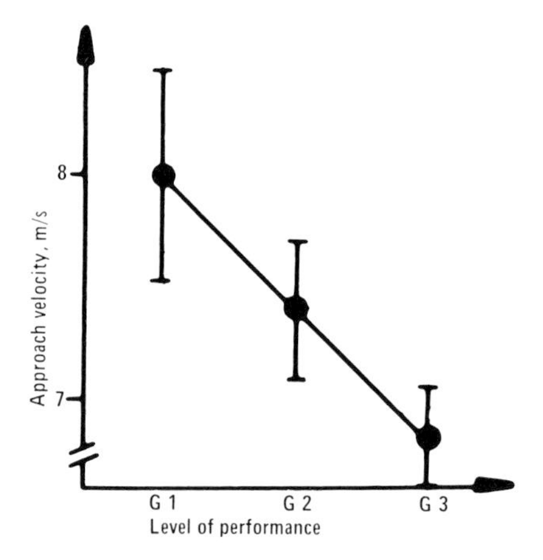

Fig. 9. Mean and SE of run-up velocities at different level of performance, group G 1 being world-class gymnasts ($n = 10$), G 2 national-class gymnasts ($n = 10$) and G 3 junior-class gymnasts ($n = 10$).

The mean of the best group (7.98 m/s) corresponds with the results (the mean being equal to 7.75 m/s) of the study of Dillmann et al. [1985]. The highest measured value from our most recent study is 8.4 m/s for a handspring with 1/1 somersault in the post-flight phase.

The Take-Off

During take-off the horizontal velocity of the centre of mass decreases while both vertical velocity and angular momentum about the transverse axis increases. The magnitude of changes in the linear and angular momenta are related to the level of performance. In contrast with the nonsignificant duration of support on the board, the mean horizontal and the vertical ground reaction forces increase with the standard of vaulting.

The angular momentum at take-off differs significantly between groups, increasing with skill level. To facilitate comparisons between subjects, angular momentum is normalized by dividing it by the moment of inertia. The resultant angular velocity of the body increases during board contact from approximately zero to 6.3 rad/s. This positive trend is further supported by the high correlation coefficient $r = 0.75$ between the take-off angular velocity and the height of the post-flight. The mean angular velocity at take-off for the best group is 6.8 rad/s. The maximal measured values [unpubl. data] for a handspring with 1/1 somersault in layout-position and an excellent Cuervo vault ar 8.9 and 8.8 rad/s, respectively.

The Pre-Flight

The pre-flight duration is extremely short in highly skilled vaults. The correlation between time of flight and level of performance (height of the post-flight) is negative ($r = -0.75$). Contrary to some coaching manuals, the data show that the gymnast's centre of mass has a positive (upward) vertical velocity at touch-down on the horse. The flight time of the group G1 (fig. 9) shows a mean of 0.20 s. Nowadays a tendency to shorten pre-flight time is developing which appears to be related to increases in pre-flight angular momentum. The most important limitation for this development is the touch-down angle on the horse ($\sim 30°$). If the touch-down angle is too small the lever arm of the gravitational force, which reduces angular momentum during the support phase [Dainis, 1981] increases. The resultant decrease in angular momentum does not allow the necessary rotation to occur in the post-flight. From the practical point of view it must be concluded that the shortening of pre-flight time may improve the height of post-flight but it must be linked with an increase in angular velocity at take-off from the board.

The Support

The duration of support seems to be an important indicant of the successful vault for it correlates well with the height of post-flight. The mean support time of the best group is 0.185 s which corresponds closely with the 0.180 s reported by Dillman et al. [1985].

During the support phase all gymnasts reduce horizontal velocity and increase vertical velocity. These changes in vertical velocity at support significantly reflect the levels of skill of the gymnasts. Angular velocity, as is to be expected, also decreases at support phase. The decrease, however, is quite marked (60%) and different from that observed in other vaults. Furthermore, there is also a decrease, though of lesser magnitude, in transverse rotation. This latter finding is at variance with the consensus found in books on coaching. Only two examples have been measured that show an increasing, although small, angular velocity. One, a handspring with 1/1 layout somersault (male gymnast); the other, a layout Tsukahara (female gymnast).

The Post-Flight

Height and distance of post-flight differentiate levels of performance. These variables, together with angular momentum, are a direct consequence of the kinetics of the preceding phases of the vault.

In table II a summary is provided of the most important data reported by different investigators for vaults of the handspring family. Cheetham [1982] and Dillman et al. [1985] analysed the voluntary handspring vault (that is handsprings with 1/1 somersault or Cuervo vaults). Our study investigated the compulsory vault (ordinary handspring) in the competitive period 1976–1980.

For comparison between the handspring vault and the Tsukahara vault the most important data are summarized in table III [Dillman et al., 1985] from which table it can be seen that differences occur between pre-flight time and between support duration. In the Tsukahara vault the time separating the first and the second hand touch-downs is relatively long. During the period only low reaction forces operate to decrease horizontal velocity and angular momentum and to increase vertical velocity. Therefore, the resultant reduction of horizontal velocity during the support phase is less in the Tsukahara vault than in the handspring vault. The geometry of take-off is similar in both techniques.

Because of the early first-hand contact in the Tsukahara, the touch-down angle on the horse is much less than that reported for the handspring.

The data suggest that there are no basic differences between the princi-

Table II. A summary of three major studies of the kinematics of the men's front handspring vault

	Cheetham [1982]	Dillman et al. [1985]	Brüggemann and Nissinen [1981]		
	college gymnasts (n = 8)	Olympic performance (n = 8)	world class (n = 10)	national class (n = 10)	junior class (n = 10)
Time, s					
Time of hurdle	–	0.23	–	–	–
Time on board	0.11	0.11	0.11	0.10	0.10
Time of pre-flight	0.15	0.17	0.200	0.228	0.280
Time on horse	0.13	0.18	0.185	0.196	0.209
Time of post-flight	1.04	0.92	–	–	–
Displacement, m					
Hurdle	–	2.45	–	–	–
Pre-flight	1.65	1.78	–	–	–
Post-flight	3.51	4.07	–	–	–
Body position, °					
Departure from board	73.6	74.9	74.3	75.9	79.7
Contact with horse	33.0	33.4	29.4	31.7	39.9
Horizontal velocity, m/s					
Board contact	7.32	7.79	7.98	7.40	6.79
Departure from board	4.53	5.11	5.17	5.24	4.70
Departure from horse	3.56	3.57	3.82	3.73	3.19
Vertical velocity, m/s					
Departure from board	4.02	4.49	4.52	4.31	3.88
Contact with horse	2.50	2.75	2.53	1.81	1.07
Departure from horse	2.86	2.97	2.75	2.37	1.06
Angular velocity, rad/s					
Departure from board	–	–	6.80	6.25	5.93
Departure from horse	–	–	3.62	3.49	3.60

Table III. Comparison of the most important data of men's handspring vs. Tsukahara vaults [Dillman et al., 1985]

	Handspring	Tsukahara
Time, s		
Time of hurdle	0.23	0.23
Time on board	0.11	0.11
Time of pre-flight	0.17	0.09/0.12[1]
Time on horse	0.18	0.28
Time of post-flight	0.92	0.86
Displacement, m		
Hurdle	2.45	2.29
Pre-flight	1.78	1.36
Post-flight[2]	3.14	3.00
Body position, °		
Board contact	117.5	117.5
Departure from board	74.9	76.1
Contact with horse	33.4	9.6
Horizontal velocity, m/s		
Board contact	7.79	7.71
Departure form board	5.11	5.21
Departure from horse	3.54	3.75
Vertical velocity, m/s		
Departure from board	4.49	4.42
Contact with horse	2.75	3.42
Departure from horse	2.97	2.54

[1] Time board to first handcontact/time between hand contact.
[2] Length of post-flight from and of the horse.

ples underlying both classes of vault. Only few data about rotation in vaulting are available in the literature and these are indicated in tables II and III. Table IV summarises information obtained from recent unpublished investigations about vaults of different difficulty and class. The table includes the handspring and the Tsukahara group performed by male and female gymnasts. The number of gymnasts analysed is given in parentheses. Angular velocity of the whole body about the transverse axis through the centre of mass at take-off from board and horse, and the change of angular velocity is presented. In comparison with the ordinary handspring, the more difficult vaults present higher values after the support phase on the horse. The higher angular velocities result from higher take-off values from the board as well as from smaller reductions during the support on the horse.

Table IV. Comparison of the angular velocity data about the transverse axis of top international gymnasts

	Departure from board, rad/s	Departure from horse, rad/s	Change during support, rad/s
Group: Handspring			
Male			
Ordinary Handspring (n = 10)	6.80	3.62	–3.18
Handspring 1/1 (n = 1) somersault 1/2 twist	7.77	6.00	–1.77
Handspring 1/1 somersault layout (n = 1)	8.80	8.90	+0.10
Cuervo (n = 2)	8.90	5.30	–3.60
Female			
Cuervo (n = 1)	6.6	6.1	–0.50
Group: Tsukahara			
Male			
Tsukahara piked (n = 10)	6.3	4.8	–1.5
Female			
Tsukahara layout (n = 1)	7.8	8.3	+0.5
Group: Roundoff entry			
Female			
backward somersault layout, 2 twists (n = 1)	10.2	10.0	–0.1

The handspring with 1/1 somersault in layout-position performed by Gousev (URS) has no angular velocity reduction during the support on the horse. The kinematics of this specific technique is illustrated by figure 10. In addition to the constant of barely increasing angular velocity, this gymnast decreases his horizontal velocity during support on the horse by only 0.1 m/s. At the same time the vertical velocity decreases, which is in contrast to all other analysed vaults of highly skilled gymnasts.

The Tsukahara vaults have lower losses of angular velocity during contact than the other vaults, a finding that should be seen in context with the lower height of the centre of mass and the longer distance of the post-flight, as described above.

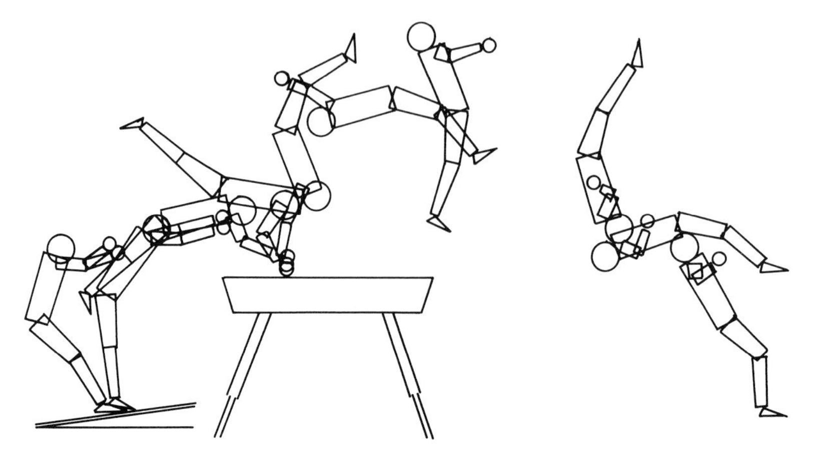

Fig. 10. Handspring 1/1 somersault in layout-position [Nissinen, internal report, 1985].

The Landing

This is discussed as one element in the next section on the mechanical loading of the body in gymnastics.

Mechanical Loads Imposed on the Body by Gymnastics

In gymnastics those moments that most usually produce the highest overload on body structures include impacts and decelerations during forceful take-offs and landings. [Nigg et al., 1982] and wrenches during swings on the still rings [Brüggemann and Nissinen, 1982]. It is to these issues that the following section is devoted. In particular, attention is directed at the following: – loads on the Achilles tendon during floor take-off; – trans-articular forces at the knee and ankle during floor take-off; – loads on the knee and hip during landing from apparatus; – trans-articular forces at the shoulder joint during the swing on stationary rings

Loads on the Achilles Tendon during Take-Off

Recent observations have shown that various pathological problems in the gymnast's feet and ankles develop in response to the stress imposed by multiple performances, yet few other than Heger [1966] and Nigg et al. [1982] have made any attempt to investigate the effects of the multiple intense loadings imposed by gymnastics on the body of the gymnast. Even Heger's study of the Achilles tendon force had its shortcomings, for the forces assumed were not based on biomechanical data and an erroneous view was taken that the point of application of the forces was invariant.

In a study conducted by Brüggemann [1984], the magnitudes of the loadings on the Achilles tendons of 4 international gymnasts during floor take-off after a round-off or flic-flac approach to a single and double back somersault, were calculated during a series of laboratory trials. In each trial the ground reaction force on the right foot, its point of application and the moment about the vertical axis, were monitored continuously using a Kistler force platform covered with a gymnastic mat. Body configuration changes were recorded on film at a nominal 200 Hz (205 and 203 Hz calibrated) using two orthogonally placed Locam cameras. For the purpose of analysis, The gymnast's legs were treated as a three-link system.

The load of muscles and tendons crossing the ankle joint can be represented by the force moments acting in three planes perpendicular to each other in the ankle joint. This model was introduced by Paul [1965] and used by Capozzo and Pedotti [1975], Groh and Baumann [1976] for the analysis of human gait and again by Baumann [1981] in a study of sprint running. The force moments at the ankle joint were computed in the three planes (xz, yz, xy) taking into account ground reaction forces and initial forces of the foot as well as their moments.

The spatial force moment vector at the ankle joint with the components M_z^A, M_y^A and M_z^A was then projected in the tibia system (fig. 11) M_z^T was the

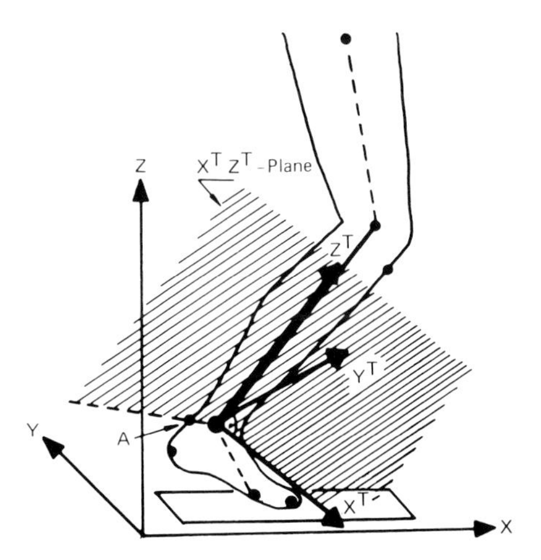

Fig. 11. Definition of the coordinate systems.

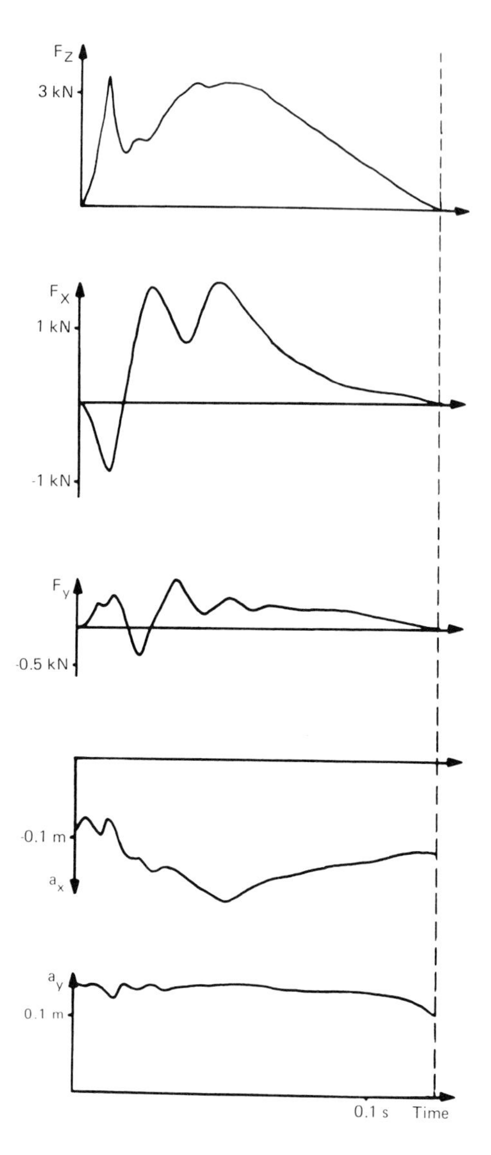

Fig. 12. Ground reaction forces (F_z, F_x, F_y) and point of application of the force vector (a_x, a_y) for a typical backward somersault take-off.

force moment acting in a plane perpendicular to the long axis of the tibia while M_y^T was the moment in a plane defined by the long axis of the tibia and the Achilles tendon represented by the markers fixed on the tendon. Then the net muscle moments were calculated from the force moments in the tibia system.

In figure 12a sample is shown of the ground reaction forces recorded from one of the subjects. After allowing for the differences in magnitude – the one foot reaction forces are about half those of the two foot forces – the patterns of forces from the right foot differ only slightly from those recorded from both feet simultaneously [Brüggemann, 1983b]. Maximal vertical forces reach 3,350 N, horizontal ones (posterior-anterior) about 1,500 N and the lateral ones up to 1,550 N. In order to compare subjects, maximum values may be normalized by dividing them by the subject's body weight. The magnitude of the vertical forces shows a range from 3.4 to 5.6 times the body weight and the horizontal reaction forces from 0.6 to 2.6 body weight. The lateral forces (F_y) seem to vary considerably among the analysed trials which is probably due to the non-perfect alignment of the gymnast at take-off.

Force vectors, shown in figure 13 with their points of applications defined, illustrate the positive values of the force moments M_x and M_y at the ankle joint during take-off. The time histories of the components of the force moments acting in the xz, yz and xy-planes at the ankle joint are plotted in figure 14. M_y^A acting in the primary plane of motion provides the major contribution.

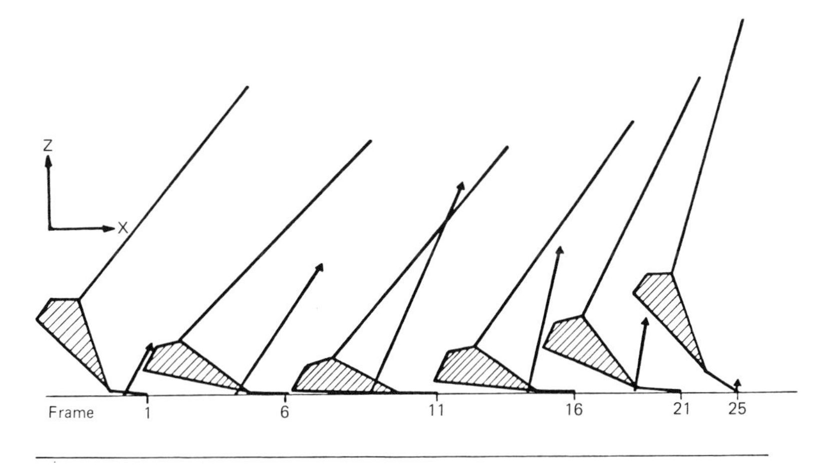

Fig. 13. Vector diagram of a typical backward somersault take-off.

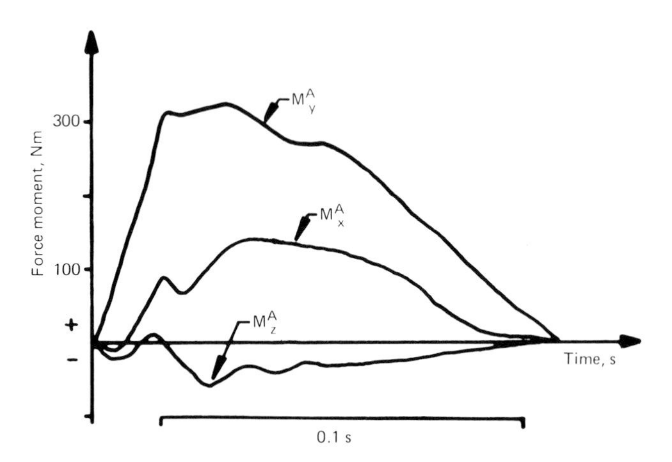

Fig. 14. Force moments of the right ankle joint during a backward somersault take-off.

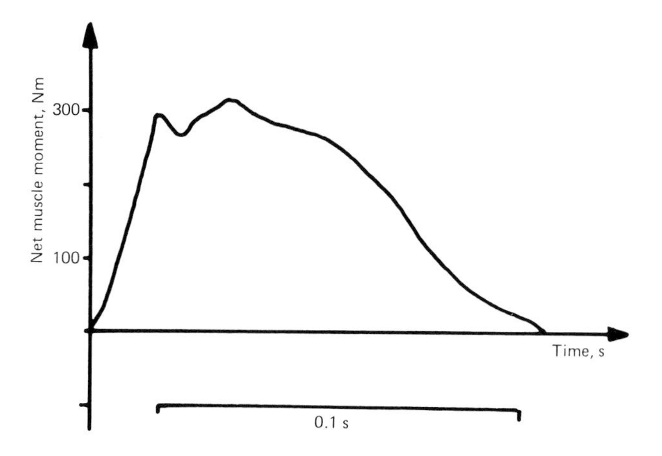

Fig. 15. Net muscle moment of the m. triceps surae during take-off in a backward somersault.

The maximum values of the force moments indicate that the ranges of M_x^A and M_z^A are very high. Those results depend on the different foot positions and different anatomical orientations of the segments with regard to the xy-plane during take-off. No subjects show a foot orientation parallel to the x-axis of the force platform. Therefore it is not only the M_y^A moment but also the M_x^A and M_z^A components which may be loading the plantar flexors. This fact is illustrated in figure 15 where the net muscle moment of the m. triceps surae transmitted by the Achilles tendon M_y^T – assuming that the activity of the dorsiflexors is negligible and that the whole moment is transmitted by the Achilles tendon – is plotted against time during the take-off period.

The assumption of minimal activity of the m. tibialis anterior was judged to be tenable from EMG measurements. Figure 16 shows the raw EMG traces of the m. triceps surae and the m. tibialis anterior as well as the vertical ground reaction force.

Figure 17 shows the force time curves of the resulting computed Achilles tendon force of all subjects. The maximum values approach 10 kN and may have been higher had the take-off for the back somersault not been constrained by the need to land on the force platform. These maximum forces represent loads about 16 times body weight.

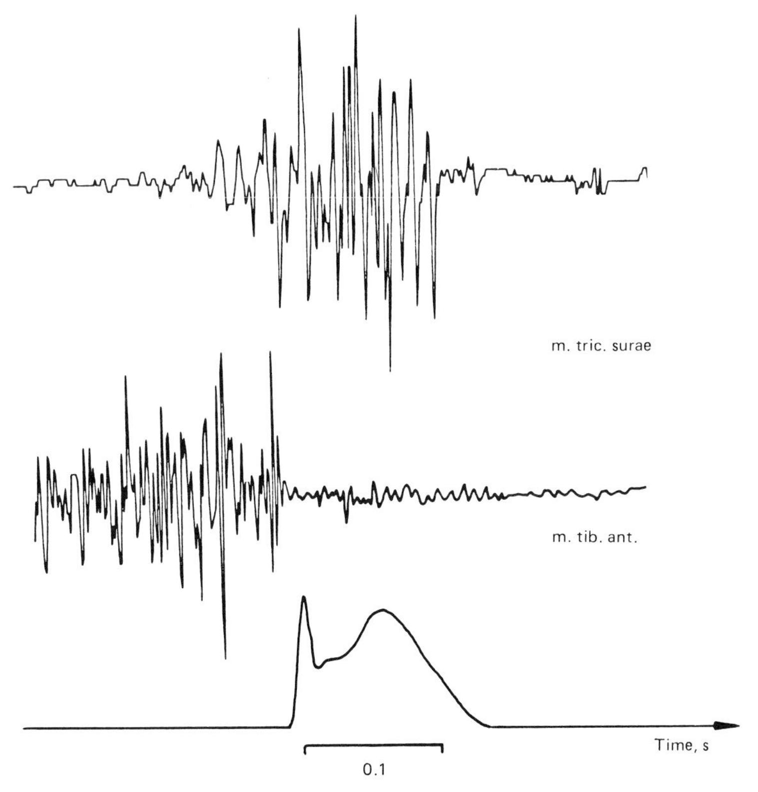

Fig. 16. Synchronization of the raw EMG of the m. triceps surae and m. tibialis anterior, and the time curve of the vertical ground reaction force during a backward somersault take-off.

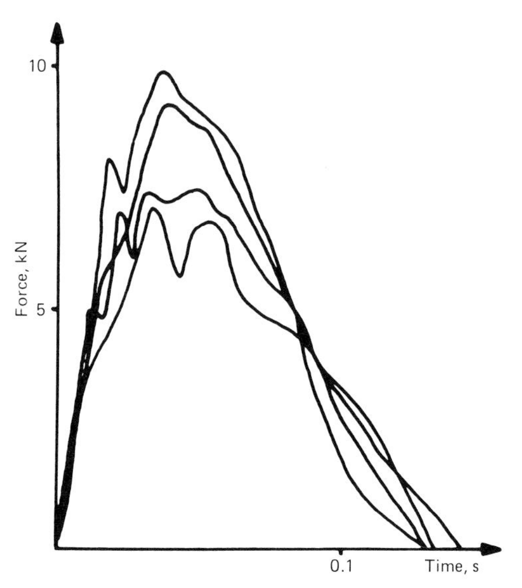

Fig. 17. Calculated Achilles tendon forces during a backward somersault take-off of 4 subjects.

With the help of computer tomography the cross-section of the Achilles tendon was measured. Mean and standard deviation were calculated at 89.1 ± 10.4 mm². Therefore, assuming a cross-section of 90 mm², the highest computed force in the Achilles tendon during take-off of the backward somersault produces a tensile stress of about 110 N/mm² (subject 1). Yamada [1970] found for the Achilles tendon an ultimate tensile strength of 55 N/mm² and Wilhelm [1972] reported an ultimate force of 9,125 N. Our calculations point out that the Achilles tendon seems to be able to tolerate higher loads during short periods.

The lever arm of the Achilles tendon computed frame by frame from the film data, ranges from 0.031 m to 0.054 m. The lever arm of one subject's foot was calculated by computerized tomography at defined ankle angles. As a result the film data calculation undervalued the lever arm by approximately 5%. Hence it follows that the Achilles tendon force may be overvalued by about 5%.

The results of this study indicate that take-off of the backward somersault has a very high loading effect on the Achilles tendon during a short period of about 100 ms.

The tendon forces seem to be very close to the earlier reported ultimate tensile strength.

In view of the capacity of the tendon to adapt slowly to high loads, high loading movements on the floor must be introduced progressively over periods of months or even years.

Trans-Articular Forces at the Knee during Take-Off

The trans-articular forces at the knee as well as at the ankle, are presumably particularly high during take-offs and landings and it is understandable that numerous reports of knee pain are to be found from gymnasts [Klümper, personal commun., 1986; Brüggemann, unpubl. report, 1986]. This is the case, for instance, during the running forward somersault on the floor. In an attempt to determine the magnitude of the knee and ankle forces in this event, 8 gymnasts were examined. Ground reaction forces invoked by both legs were monitored continuously using a Kistler force platform. In this experiment the trials were conducted on the platform without any damping mat. Therefore the registered peak forces will exceed those arising during performance on a regular gym floor. The movements were filmed simultaneously by a Locam camera operating at 200 fps as described earlier.

For the purpose of analysis, equivalent loadings were assumed on both legs, and the legs were treated as a three-link-system of rigid bodies. From the joint reaction forces and the active muscle forces, the bone-on-bone forces were calculated in the knee and ankle joint [Winter, 1979]. The divi-

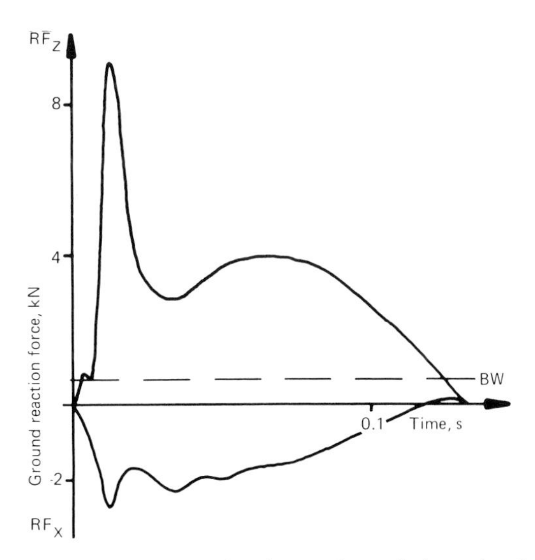

Fig. 18. Ground reaction forces of a typical running forward somersault take-off. The dotted horizontal line represents the subjects's body weight (BW).

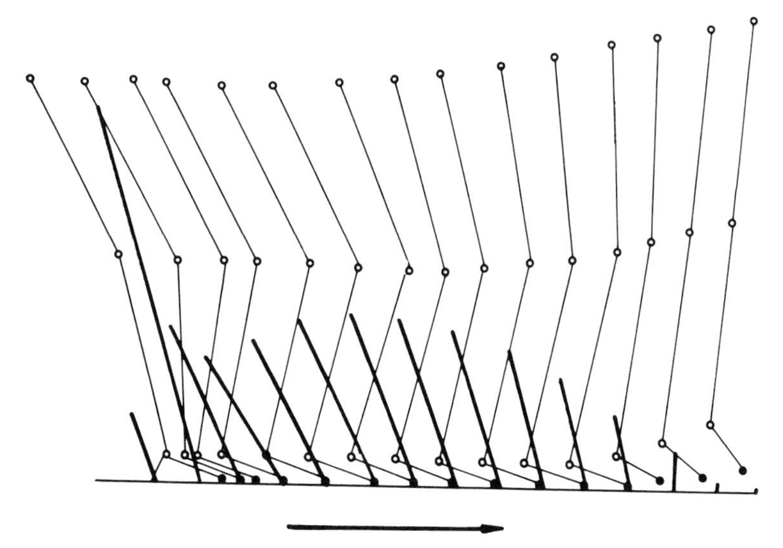

Fig. 19. Vector diagram of a running forward somersault take-off.

sion into compressive and shearing forces occurred according to the spatial orientation of the tibia system. The pattern of ground reaction forces plotted against time (fig. 18) were found to be somewhat similar to those recorded during long jump or high jump take-off [Nigg, 1974]. The vertical and horizontal components of the reaction vector had peaks of 9,000 and 4,000 N, respectively.

The time of take-off varied from 100 to 150 ms. Assuming that the point of application of the forces was located in the region of os metatarsale V, figure 19 illustrates the force vector diagram for a typical subject. Joint reaction forces and net muscle moments were calculated from the kinetic and kinematic raw data. Assuming a constant lever arm for the patella tendon, the net muscle force acting upon the knee joint was analysed. The bone-on-bone force of the knee joint is plotted against time in figure 20. The resulting forces indicate, that take-off of the forward somersault has a high loading effect on the knee joint. Assuming a congruent surface of the femuro-tibial joint of about 4 cm^2 [Nigg et al., 1982], the pressure will be estimated as 2,250 N/cm^2. Yamada [1970] reported a critical pressure of 500 N/cm^2 for cartilage. These results support the view [Viidik, 1980] that the knee joint in vivo is able to tolerate higher dynamic than static pressures.

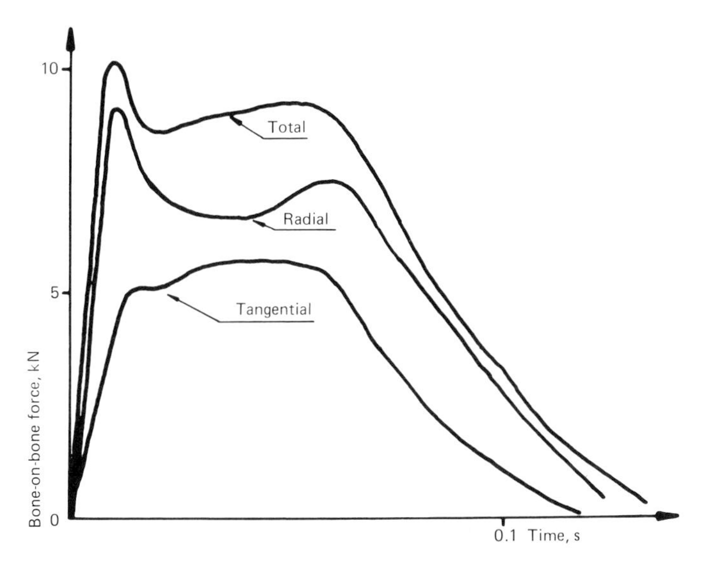

Fig. 20. Bone-on-bone force of the knee joint during a typical running forward somersault take-off.

Load during Landing from the Apparatus

One moment of high stress in gymnastics occurs during dismounts and landings where, despite the use of cushioning mats of ever improving quality, reports of back pain continue to be received [Nigg et al., 1982].

Nigg and Spirig [1976], for instance, attempted to estimate the magnitude of these loading forces at touch-down by analysing the vertical peak accelerations of the tibia, hip and head during landing under different conditions. They fixed three accelerometers to the tibia, hip and head and telemetred the accelerations at a nearby recorder during dismounts and landings from the apparatus.

Peak accelerations of 30 g for the tibia and 10 g for the hip were recorded by Nigg et al., [1982] in these landings from heights of 1.5 m. The peaks were observed to vary with the landing technique adapted and, in particular, from the damping effect of the leg movement adopted by different individuals. The quotient of hip and tibia acceleration was used to quantify the individual landing technique.

Analysing members of the Swiss national team, the leg damping quotient varied from 0.28 to 0.52. This result underlines the important role of the landing movement in reducing the load on the spine [Nigg, 1974].

The influence of the landing surface on the transmission of impact forces was evaluated by comparing similar dismounts on two different surfaces, one a normal mat the other and enhanced shock-absorbent surface. The effect of the new mat was to reduce accelerations at the tibia from 50 to 10 g and at the hip from 20 to 8 g [Nigg, 1974]. This provides a clear indication of the importance of the landing surface in reducing stresses on the lower extremities and, therefore, also on the spine. As gymnastic performances improve and landing techniques increase in complexity, the need for safe landing surfaces will become ever more pressing.

Trans-Articular Forces at the Shoulder Joint

Technique on the still rings has developed greatly in the last 15 years. Nowadays, a technically superior performance of the giant swing is characterized by a straight arm execution, keeping the arms and rings parallel during the downward swing. The increases in swing amplitude combined with associated increases in mechanical energy result in an increased incidence of pain in the shoulder joint.

While the reaction forces experienced during exercises on the rings have been measured by Marhold [1961] during a study of equipment safety, and by Sale and Judd [1974], the only studies of joint loads evoked during dynamic movements on the stationary rings, are those by Nissinen [1983] and Ballreich, Brüggemann and Nissinen [unpubl. report, Bundesinstitut für Sportwissenschaften, 1981]. In the latter two studies reaction forces on the rings were recorded using two piezo-electric force transducers. A lateral view of the performances was simultaneously filmed with a Locam camera operating at 100 fps. From the coordinates of the segment endpoints, the reaction forces and the individual anthropometric characteristics, the net muscle moments and the joint reaction forces were calculated.

It was found that the maximum mechanical load occurs when the gymnast approaches vertical body alignment at the end of the downward swing. At this moment the hip and shoulder joints are nearly extended. Estimates of the joint reaction forces at the shoulder at this moment, were calculated and are plotted in figure 21. The moment of vertical body alignment is marked by the vertical dotted line. The vertical joint reaction forces of the shoulder and hip joints increase during the downward swing while the horizontal forces decrease. After vertical body alignment, the horizontal force changes direction and increases in the opposite direction. This change is combined with a reduction of vertical joint reaction force. Maximum values of 2,100 and 1,500 N for the shoulder and hip joint, respectively, have been registered. An

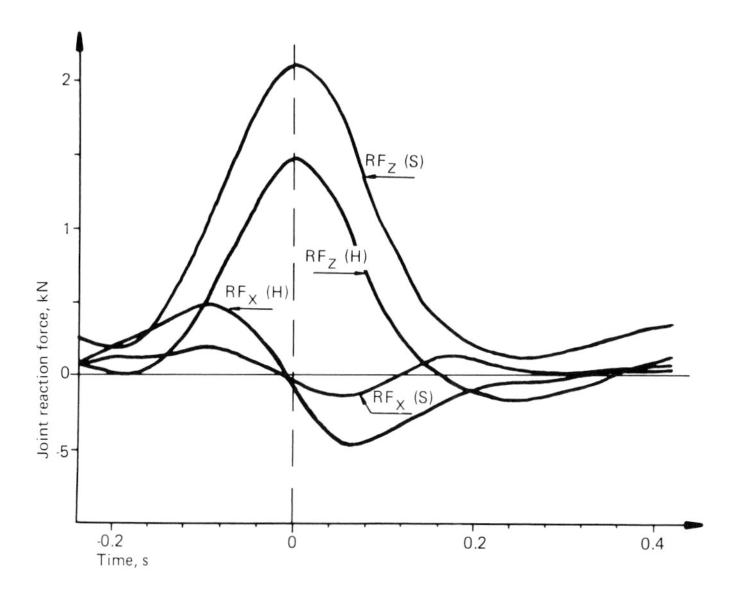

Fig. 21. Joint reaction forces of the hip (H) and shoulder (S) joint during the most important phase of the giant swing on the rings. The dotted line indicates the moment when the body of the gymnast is aligned vertically.

analysis of gymnasts with a lower level of technical skill shows, somewhat surprisingly, that their maximum joint reaction forces are even higher than those of the highly skilled.

The muscles acting at the shoulder do not work as initiators or prime movers in the giant swing but as stabilisers, protecting – in collaboration with the joint ligaments – the integrity of the shoulder joint during the severe cyclic loadings the activity imposes on its structures. No comparable loadings on this joint can be found in the literature. Consequently, if injuries are to be avoided, one hardly needs to stress the importance of ensuring that the gymnast is adequately prepared both in technical skill and physical strength and power.

Conclusion

There is no doubt that the practice of gymnastics is producing a considerable load on the musculo-skeletal system. The major factors that are able to influence the magnitude of load seem to be the equipment used for landing and the movement technique itself.

The high loading of the Achilles tendon during the take-off of a backward somersault, of the knee joint during the running forward take-off, of the shoulder joint during giant swings and dislocates on the rings, necessitates a long, progressive conditioning of the gymnast and an adaptation of the biological structures to high loads.

From the view of motor learning as well as from the view of loads accompanying the execution of gymnastic skills, training and preparation of Olympic gymnasts requires time and thoughtful planning.

References

Bajin, B.: Goniometric analysis of the pushoff phase during the 1½ somersault in men's gymnastic vaulting; in Therauds, Science in gymnastics, pp. 1–8 (Academic Press, Del Mar 1979).

Baumann, W.: On mechanical load on the human body during sport activities; in Morecki, Fidelius, Biomechanics VII-B, pp. 79–87 (University Park Press, Baltimore 1981).

Brüggemann, P.: Biomechanical analysis of selected vaults on the longhorse; in Therauds, Science in gymnastics, pp. 9–24 (Academic Press, Del Mar 1979).

Brüggemann, P.: Biomechanische Analyse symmetrischer Absprungbewegungen im Gerätturnen (Bartels & Wernitz, Berlin 1983a).

Brüggemann, P.: Kinematics and kinetics of the backward somersault take-off from the floor; in Matsui, Kobayashi, Biomechanics VIII-B, pp. 793–800 (Human Kinetic Publishers, Champaign 1983b).

Brüggemann, P.: Mechanical load on the Achilles tendon during rapid dynamic sport movements; in Perren, Schneider, Biomechanics: current interdisciplinary research, pp. 669–674 (Nyhoff, Dordrecht 1985).

Brüggemann, P.; Nissinen, M: Analyse kinematischer Merkmale des Handstützüberschlages beim Pferdsprung (Kinematics of the handspring vault). Leistungssport *11:* 537–547 (1981).

Brüggemann, P.; Nissinen, M.: Geräteoptimierung im Kunstturnen aus der Sicht der Biomechanik; in Göhner, Verletzungsrisiken und Belastungen im Kunstturnen, pp. 100–113 (Hofmann, Schorndorf 1982).

Capozzo, L. T.; Pedotti, A.: A general computing method for the analyses of human locomotion. J. Biomech. *8:* 307–320 (1975).

Carraso, R.: Essai de systematique d'enseignement de la gymnastique aux agres (Vigot, Paris 1972).

Cheetham, P. J.: The men's handspring front one and a half somersault vault. Relationship of the early phase to post-flight; in Therauds, Biomechanics in sport, pp. 231–247 (Research Center of Sports, Del Mar 1982).

Clauser, C. E.; McConville, J. T.; Young, J. W.: Weight, volume and center of mass of segments of the human body. (AMR L-TR-69-70, Wright-Patterson Air Force Base, Ohio 1969).

Dainis, A.: Cinematographic analysis of the handspring vault. Res. Q. *50:* 341–349 (1979).

Dainis, A.: A model for gymnastic vaulting. Med. Sci. Sports Exerc. *13:* 34–43 (1981).
Dillman, C. J.; Cheetham, P. J.; Smith, S. L.: A kinematic analysis of men's olympic long horse vaulting. Int. J. Sport Biomech. *1:* 96–110 (1985).
Djatschkow, W. M.: Die Steuerung und Optimierung des Trainingsprozesses (Bartels & Wernitz, Berlin 1974).
Dusenbury, J. S.: A kinetic comparison of forward and reverse giant swing on the still rings as performed by gymnasts with varying body type; MS thesis, University of Massachusetts (1968).
Feller, I.: Absprünge rückwärts im Kunstturnen (Backward somersault take-off in gymnastics); Diplomarbeit am Laboratorium für Biomechanik der ETH Zürich (1975).
Fetz, F.; Opavski, P.: Biomechanik des Gerätturnens (Limpert, Frankfurt 1968).
Groh, H.; Baumann, W.: Joint and muscle forces acting in the leg during gait; in Komi, Biomechanics V-A, pp. 328–333 (University Park Press, Baltimore 1976).
Hanavan, E. P.: A mathematical model of the human body (AMRL-TR-64-102, Wright-Patterson Air Force Base, Ohio 1964).
Hay, J. G.; Wilson, B. D.; Dapena, J.: Woodworth, G. O.: A computational technique to determine the angular momentum of a human body. J. Biomech. *10:* 269–277 (1977).
Heger, H.: Verletzung und Belastung der Achillessehne des Geräteturners. Prax. Leibesübungen *1966:* 27–29.
Leirich, J.; Rieling, K.: Zur strukturellen Anordnung der Übungen des Gerätturnens. Die Sprungbewegungen. Theor. Prax. Körperkult. *2:* 139–147 (1967).
Leirich, J.; Rieling, K.: Zur strukturellen Anordnung der Übungen des Gerätturnens. Die Überschlagbewegungen. Theor. Prax. Körperkult. *12:* 1073–1084 (1968).
Marhold, G.: Über die Belastungsgrößen bei Übungen an den Ringen. Theor. Prax. Körperkult. *4:* 439–444 (1961).
Nelson, R. C.; Gross, T. S.; Street, G. M.: Vaults performed by female olympic gymnasts. A biomechanical profile. Int. J. Sports Biomech. *1:* 111–121 (1985).
Nigg, B. M.: Sprung, Springen, Sprünge (Juris, Zürich 1974).
Nigg, B. M.; Denoth, J.; Unold, E.: Belastungen des menschlichen Bewegungsapparates bei ausgewählten Bewegungen des Gerätturnens; in Göhner, Verletzungsrisiken und Belastungen im Kunstturnen, pp. 20–38 (Hofmann, Schorndorf 1982).
Nigg, B. M.; Spirig, J.: Erschütterungsmessungen beim Kunstturnen. Leistungssport *2:* 91–96 (1976).
Nissinen, M.: Kinematic and kinetic analysis of the giant swing on rings; in Matsui, Kobayashi, Biomechanics VIII-B, pp. 781–786 (Human Kinetic Publishers, Champaign 1983).
Paul, J. P.: Bioengineering studies of the forces transmitted by joints; in Kenedy, Biomechanics and related bioengineering topics, pp. 369–380 (Pergamon Press, Oxford 1965).
Payne, A. H.; Barker, P.: Comparison of the take-off forces in the flic-flac and the back somersault in gymnastics; in Komi, Biomechanics V-B, pp. 314–321 (University Park Press, Baltimore 1976).
Peek, R. W.: A cinematographical and mechanical analysis of the straight arm backward giant swing on the still rings; MS thesis, Springfield College (1968).
Sale, G. D.; Judd, L. R.: Dynamometric instrumentation of the rings for analysis of gymnastics movements. Med. Sci. Sports Exerc. *6:* 209–216 (1974).
Thevenin, J.: Gymnastique – vers un langage commun. Educ. phys. Sport *157:* 64–68 (1979).
Viidik, A.: Elastomechanik biologischer Gewebe; in Cotta, Krahl, Steinbrück, Die Belastungstoleranz des Bewegungsapparates, pp. 124–135 (Thieme, Stuttgart 1980).

Wilhelm, K.: Die statische und dynamische Belastbarkeit der Achillessehne. Res. exp. Med. *132:* 157–159 (1972).
Winter, D. A.: Biomechanics of human movement (Wiley, New York 1979).
Yamada, H.: Strength of biological material (Huntington, New York 1970).

Dr. P. Brüggemann, Deutsche Sporthochschule Köln, Postfach 450327,
D-5000 Köln (FRG)

Med. Sport Sci., vol. 25, pp. 177–194 (Karger, Basel 1987)

EMG Quantification and Its Application to the Analysis of Human Movements

M. Van Leemputte, E.J. Willems
Instituut voor Lichamelijke Opleiding, Katholieke Universiteit Leuven, Belgium

The use and interpretation of electromyography (EMG) is extensively discussed in the literature, generated by several disciplines. Even within the field of the analysis of human movements, EMG is used for different purposes such as the examination of the initiation times of muscles, the evaluation of muscle function, the determination of the relative contribution of synergic muscles and the quantification of muscle fatigue. In this study, discussion is limited to the quantification of EMG and to the relation between the quantified EMG and the degree of muscular activity, a relation widely held to be proportional to the isometric muscle force. After all, the applications mentioned above depend largely upon this relation.

Some Reflections on EMG Quantification

Available Methods

In 1952 Inman et al. reported a parallelism between the isometric force of a cineplastic m. biceps brachii and the low-pass filtered rectified EMG. The applied filter was specified as: '...that the output should be smooth enough to permit useful analysis...' [p. 188], and in this sense EMG is still quantified. They continue: '...the circuit was called an integrator for lack of a better name' [p. 188]. This was the beginning of the 'integrated EMG' and also the start of a lot of confusion. Indeed, already in the same year, Lippold [1952] analysed the EMG by measuring the area of the rectified signal by means of a planimeter; a linear relation between this integrated EMG and the isometric force was found. In an attempt to clarify the terms, Bouisset

[1973] named the latter method a 'true integration', while Inman's method was referred to as 'continuous mean voltage', 'averaged EMG' or 'envelop value'. Also, Winter et al. [1979] argued for more uniformity and more useful technical information when reporting EMG studies.

The integrated EMG (IEMG) is now the most widely used method to estimate the degree of muscular activity. Literature about the relation between the isometric force and the IEMG is abundant and there is no doubt that a parallelism exists between them. However, the shape, the accuracy and the reproducibility of this relation are topics for discussion. Not only strong linear relations [Komi, 1973; Hagberg, 1979] or partial linear with an increased steepness at high forces [Kuroda et al., 1970; Bigland-Ritchie, 1981] but also quadratic relations [Simons and Zuniga, 1970; Rau and Vredenbregt, 1973] were found between the IEMG of m. biceps brachii derived from surface electrodes and the isometric torque at 90 degrees elbow flexion.

Beside the integration, EMG is also quantified by means of parameters derived from the frequency of the signal. According to Basmajian et al. [1975] and Sato [1982], a spectrum analysis is more useful to evaluate muscle fatigue and pathology.

Numerous other methods have been developed in order to improve the quality of the quantified EMG. These include: peak-to-peak amplitude [Knowlton et al., 1956], number of peaks [Hayward, 1977] and peaks per spike [Magora and Gonen, 1977]. Fewer applications of these techniques are to be found despite the strong linear relations with the isometric force reported.

Factors Affecting the Isometric Force-EMG Relationship

The slope of any regression line between the force and the associated quantified EMG depends upon numerous factors, such as the characteristics of the amplifier, the location and orientation of the electrodes and the inter-electrode distance [Vigreux et al., 1979], the skin impedance [Rau, 1974] and temperature [Petrofsky and Lind, 1980].

With regard to some inconsistent findings, more important are conclusions wich may affect the shape (linear, quadratic) of the relation. Møller [1966, cit. in Vigreux et al., 1979] found a more linear force-EMG relation when the inter-electrode distance was smaller. Maton and Bouisset [1977] suggested that the shape of the force-EMG relation depends on the ratio between the number of slow and fast muscle fibres. Benoit and Hainaut [1973] found a more linear relationship when values are obtained from a contraction with a gradually increasing force compared to contractions at various constant force levels. These findings were confirmed by Heckathorne and Childress [1981].

According to Sherif et al., [1983] not enough attention is paid to the effect of a time delay between the start of the electrical activity and the onset of the mechanical answer. For the electromechanical delay of m. biceps brachii, Norman and Komi [1979] reported a value of 41 ms while Corser [1974] found values between 30 and 60 ms. Particularly in anisotonic contractions this delay should be taken into account.

When the external torque at a joint results from the action of more muscles, Hof and Van Den Berg [1977] stated that an EMG analysis based on only one muscle may lead to unpredictable relations. They found a linear relation between the weighted sum of low-pass filtered EMG's of three calf muscles and the net torque at the ankle, whereas for the individual muscles these relations were unclear.

EMG Quantification in Dynamic Situations

Beside the complications discussed above, additional problems occur in dynamic situations. These problems are hard to investigate because of the lack of an available criterion. The dynamic force produced by a muscle is no longer proportional to the degree of muscular activity because other factors may affect the muscle force, such as a change of the muscle length and a change of the contraction velocity. A possible way to study the relation between the quantified EMG and the degree of muscular activity is to assume a constant, mostly maximal, degree of activity during the dynamic contraction.

Several authors agreed that the integrated EMG is not affected by a length change of the muscle at maximal effort [Komi and Buskirk, 1972; Seliger et al., 1980], while others, such as Rosentswieg and Hinson [1972], found a contrary behaviour. With regard to the contraction velocity, Barnes [1980] reported at maximal effort, a decreasing value for the IEMG with increasing velocity, which is explained in the light of 'the size principle' [Henneman and Olson, 1965]. Others reported that the IEMG at maximal effort is not affected by a change of velocity [Hinson and Rosentswieg, 1973; Komi and Rusko, 1974; Rothstein et al., 1983].

An important problem for the IEMG, particularly in dynamic situations, is a fluctuation of the baseline of the raw EMG. Thys et al. [1977] concluded that the use of a high-pass filter of 25 Hz does not affect the force-IEMG relationship and baseline fluctuations are properly reduced.

Vredenbregt and Rau [1973] suggested that an integrated EMG should be based on EMG samples of at least 5 seconds in order to obtain a reliable value. On the other hand, Grieve [1975] reported the theoretical necessity of EMG samples shorter than 50 ms in order to be useful in the analysis of human movements. Khalil et al. [1976] compared the accuracy of the IEMG of m. biceps brachii based on samples of 15, 30, 60 and 120 ms. They found a non-specified 'good correlation' with force when the samples were longer than 60 ms. Norman et al. [1978] found a strongly reduced coefficient of variation with integration times longer than 80 ms.

Summary

Efforts to quantify the relationship between the isometric force generated by the muscle and the associated electromyogram have largely focussed on the integration (IEMG). Although high linear correlations between force and IEMG have been reported in literature, the form of that relationship has been shown frequently to deviate from simple linearity. Moreover, the results of any given method of quantification are not always reliably reproducible. Reasons for some of the inconsistencies, such as the effects of the electromechanical delays or synergic muscle activity, have been mentioned.

For the relationship between dynamic muscular force and the electromyogram to be quantified, an additional constraint exists, that of sample duration. The time over which quantification is carried out, it has been shown, must be short, probably about 50 ms.

Development of a Specific Quantification Technique

Several well-known difficulties hinder the quantification of the electromyogram, and these become more serious during dynamic activity. The authors have attempted to develop a method which takes into account some of these difficulties, including the instability of the baseline (d.c. shifts), electromechanical delay, the undue dependency of quantification on the amplitude of the signal and the shortness of the sampling time that must be adopted if there is to be any sensible evaluation of dynamic activity.

The modern trend in handling the EMG following digitisation is to use software procedures for the quantification. For this reason the methods discussed below are described as calculation procedures. However, it will be appreciated that real-time hardware processing is also possible. Whichever process is undertaken, however, three basic principles are involved. These are explained in the sections that follow.

Reducing the Effects of EMG Amplitude Changes

In dynamic movements the amplitude of the raw EMG from surface electrodes may change abruptly because of extraneous events. Variations in signal strength may arise, for instance, from the stretching or the displacement of the skin leading to variations in muscle-electrode and electrode-electrode spacing and skin-electrode resistance. In order to reduce this disturbing effect in the quantification of the EMG, the original signal is normalized with reference to its peak amplitude. This is achieved as follows. Calculate the maximum peak-to-peak amplitude (A(i)) for a given (ith) period. Divide each digitized (jth) value of the EMG for that period by A(i) raised to the power a. Note that the exponent a lies between 0.0 (no transformation) and 1.0 (the obtained signal is totally independent of the amplitude of the EMG). Repeat for all n-periods of the recording and for all m-digitized points in each period. The signal is called the 'transformed EMG' (TEMG(i,j)). More precisely TEMG(i,j) is defined as follows:

$$\mathrm{TEMG}(i,j) = \mathrm{EMG}(i,j) \cdot A(i)^{-a},$$

with A(i) = ABS (MAX(EMG(i,j),j=1,m) – MIN(EMG(i,j),j=1,m)), and i=1,2,3…n with n = p/m; p = total number of points sampled in the experiment; m = number of digitized points within a sample; ABS = absolute value; a = exponent reflecting to what extent the influence of the amplitude should be lowered.

Differentiation of the EMG

In most methods of EMG quantification it is assumed as a basic premise that the force delivered by a muscle is related to the amount of electrical activity. On the contrary, the method proposed starts from the assumption that the force is proportional to the changes of the electrical activity, hence to the differential rather than to the integral of the EMG signal. Therefore this step of the method is called the 'differentiated EMG' (DEMG) which for a sample (i) equals:

$$\mathrm{DEMG}(i) = \sum_{j=1}^{m} \mathrm{ABS}\,(\mathrm{EMG}(i,j-1) - \mathrm{EMG}(i,j))/dt,$$

with dt = time interval between two digitized points.

In the same notation the IEMG of a sample (i) equals:

$$\mathrm{IEMG}(i) = \sum_{j=1}^{m} \mathrm{ABS}\,(\mathrm{EMG}(i,j) - R(i)) \cdot dt,$$

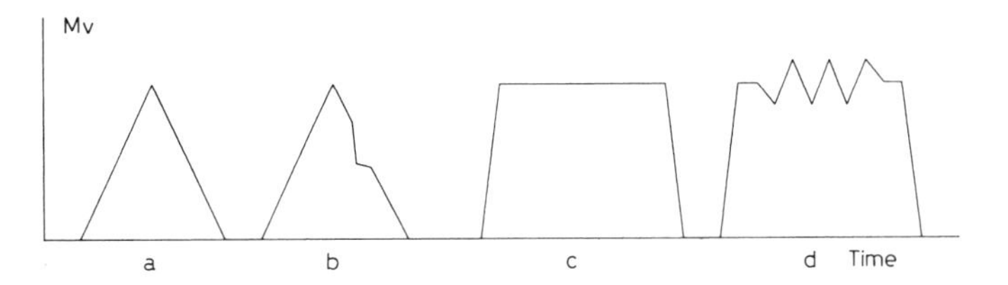

Fig. 1. Theoretical examples of EMGs. IEMG a is greater than IEMG b. For the DEMG the order is reversed. Comparing signals c and d, the IEMGs are equal and rather high, mostly depending upon the level of the signals, whereas the DEMGs of c and d are unequal, low and almost independent of the level of the signals.

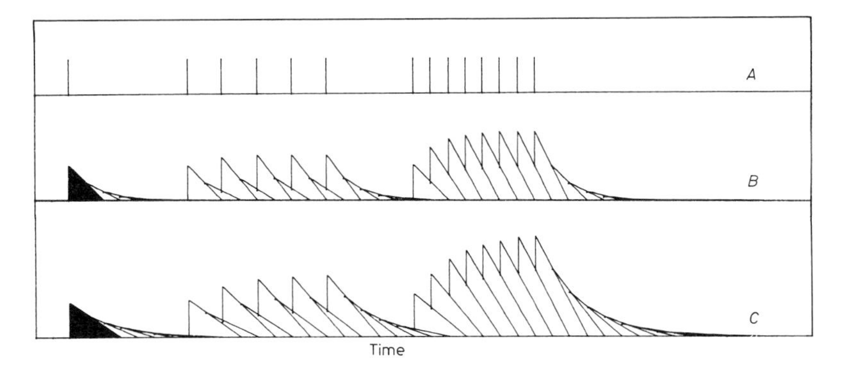

Fig. 2.A Constant amounts of electrical activity occurring at various time instants. *B, C* Resulting force build-up with a time constant of 100 *(B)* and 150 ms *(C)*.

with R(i) = level of the base line as a mean value of sample (i).

Interest in the possibilities of utilising the differentiated EMG signal was initially inspired by observations of the effects of electrostimulation – the reverse situation – for it is now widely known that a muscle reacts and generates force as a function of the changes in the applied stimulus rather than as a function of the level of the stimulus. Figure 1 illustrates some differences between the IEMG and the DEMG for four theoretical EMGs.

The Digressive After-Effect of Electrical Activity

Another basic idea is derived from the shape of a force twitch associated with an electrical stimulus (the motor unit action potential). In figure 2A a constant amount of electrical activity is assumed at various time instants. Each stimulus results in an immediate force build-up, to a level proportional to the intensity of the stimulus, followed by a linear decay. If a next twitch appears before the end of the decay, the twitch is superimposed on the remaining part of the previous one. From mathematical considerations, the linear decay is a discrete approximation of the time constant of a first order low-pass filter. Therefore, when the integrated or differentiated EMG is low-pass filtered this needs to be acknowledged by using the terms 'filtered IEMG' (FIEMG) or 'filtered DEMG' (FDEMG). The effect of a time constant of 100 and 150 ms is shown in figure 2B and 2C, respectively.

Evaluation of the 'Filtered Differences of the Transformed EMG' (FDTEMG)

Whether the three procedures described above really reflect the underlying physiological connection between force and EMG is an interesting problem. However, the main issue to be considered is whether or not a more accurate estimate of muscular activity can be achieved using the FDTEMG procedure rather than the classic method of IEMG. Therefore, the accuracy using the three procedures described above for deriving the filtered differences of the transformed EMG was evaluated and compared to the IEMG.

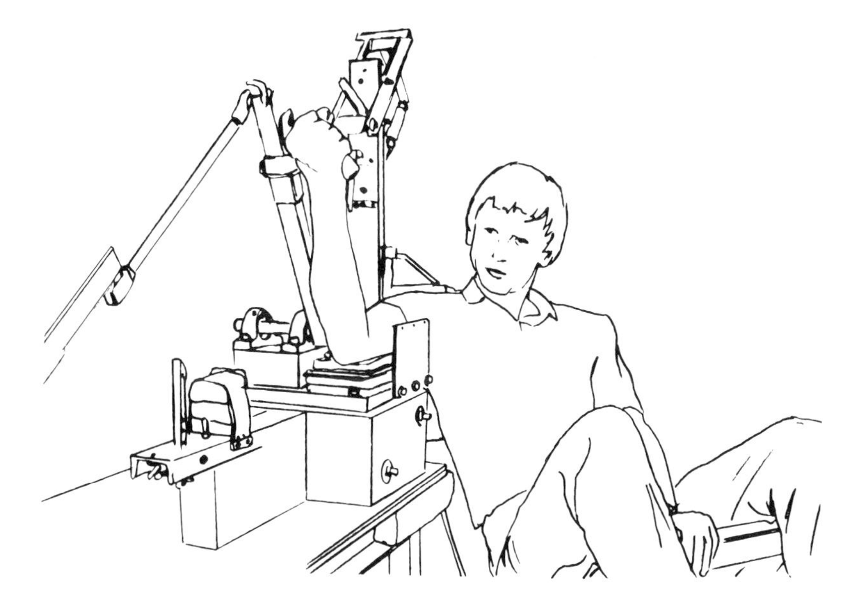

Fig. 3. Apparatus to measure the static elbow flexion torque at various angles.

Method

Forty male subjects, aged from 19 to 23 years, participated in the experiment. The subject sat on a chair with the upper-arm in an U-shaped horizontal support (fig. 3). The hand grip was connected to a torque meter (Hottinger-Baldwin T4W) at six different angles. Bipolar surface electrodes were placed longitudinally on the belly of m. biceps brachii, m. brachialis and m. brachioradialis. The subject was asked to perform a maximal isometric elbow flexion starting from rest. The test was executed at six different angles: 75, 100, 120, 140, 160 and 175 degrees. In addition, three submaximal isometric contractions were performed at intermediate angles. The exerted torque and the raw EMGs were digitized at a sample rate of 500 Hz for 3.6 s. In a retest, between 7 and 12 days later, all measurements were repeated.

The criterion used to evaluate both the IEMG and the FDTEMG was the accuracy of a mathematical relation between these signals and the static torque. From the literature it is clear that both the linear and quadratic relationships must be examined. The accuracy of the linear regression between the quantified EMG and the static torque is reported both as a coefficient of nondetermination ($1-r^2$) and also as a standard error of estimate. The standard error of estimate was also used as the criterion against which to judge the accuracy of the quadratic regression.

The effect of the electromechanical delay was evaluated by advancing the measured torque (0, 20, 40, 60 and 80 ms) with respect to the EMG. The effect of the sample time was evaluated for five different sampling durations: 20, 40, 80, 120 and 180 ms. In order to optimize the method, the accuracy was calculated for different time constants of the filter (2, 40, 80, 160 and 280 ms) and for different values of the coefficient a (0.0, 0.3, 0.4, 0.5, 0.6, 0.7, 0.8 and 1.0). For each combination the standard error of estimate of the regression relationship between the quantified EMG and the generated torque and the mean coefficient of nondetermination was calculated based on 360 contractions ($40 \cdot 9$).

Results of a Linear Regression

Both for the IEMG and for the DEMG a delay of 20 ms resulted in a minimal coefficient of nondetermination, although no significant differences were found between the 20- and the 40-ms conditions. In contrast, delays of 0, 60 and 80 ms showed significantly higher coefficients of nondetermination. The obtained optimal delay time corresponds to values referred to earlier [Corser, 1974; Norman and Komi, 1979].

Plotted as a function of sample duration in figure 4, with delay time set

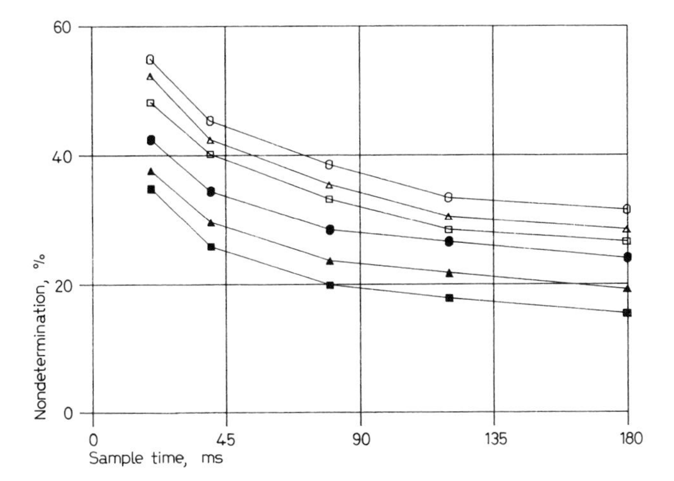

Fig. 4. Mean (n = 360) indices of nondetermination between the torque and (open symbols) the IEMG and (closed symbols) the DEMG as a function of the sample time for m. brachialis (o), m. biceps brachii (Δ) and m. brachioradialis (□).

to the optimum, are the mean coefficients of nondetermination of the 360 sampled contractions for both the IEMG and the DEMG for the three chosen muscles. From this figure it appears that integrations based on 20-ms sample times yield poor estimates of the exerted moments (a predictive efficiency of only about 50% was found). More accurate estimations were obtained when longer sampling durations are used. These findings are in agreement with Norman et al. [1978]. An evaluation of the DEMG leaded to a similar general shape of these coefficients, but the prediction accuracy of the linear relation between the DEMG and the static torque for the three muscles and for all evaluated sample times was significantly higher than the accuracy obtained with the IEMG. A mean improvement over the IEMG of 42% was found with the DEMG. It is thought that this is due to the low sensitivity shown by the DEMG to a slowly changing baseline. Figure 5 illustrates that small changes in the baseline result in important errors during integration when computing the IEMG.

The influence of the carry-over effect of successive stimuli for the DEMG (FDEMG) and for the IEMG (FIEMG) is shown in figure 6. These values were obtained with a sample time of 80 ms. Both for the FIEMG and for the FDEMG minimal coefficients of nondetermination are found with a filter time constant of 160 ms. The use of an optimal filter results in an improvement over the DEMG (zero time constant) of 63%. Its use also

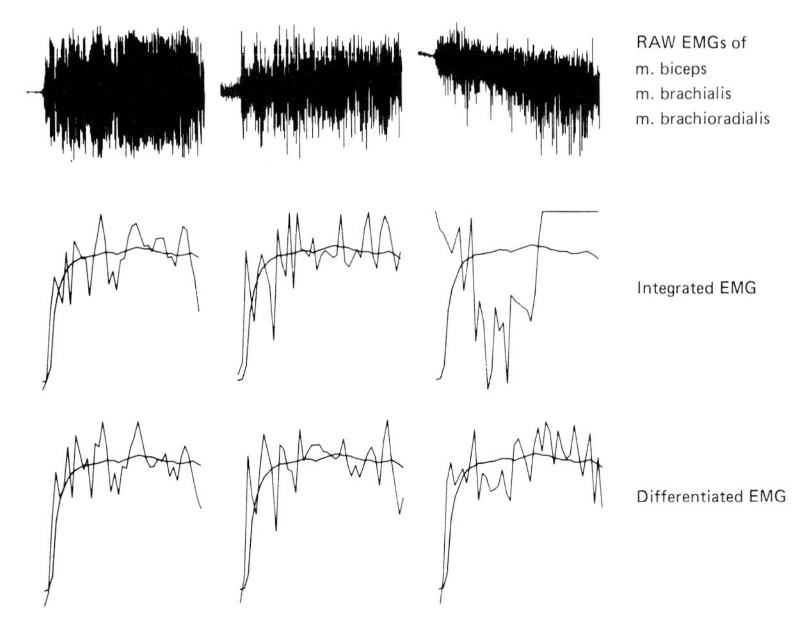

Fig. 5. The IEMGs and DEMGs of three elbow flexor muscles are shown superimposed upon a recording of the related isometric torque. The instability of the baseline of m. brachioradialis renders IEMG analysis of this muscle suspect. The contraction time was 3.6 s and the quantification procedure was based upon 80-ms samples. The smoothed line is the static elbow torque.

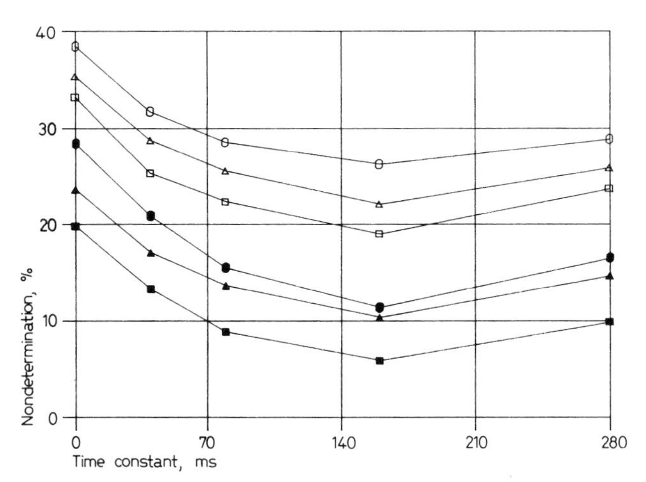

Fig. 6. Mean (n = 360) indices of nondetermination between the torque and (open symbols) the FIEMG and (closed symbols) the FDEMG as a function of the time constant of the filter for m. brachialis (○), m. biceps brachii (Δ) and m. brachioradialis (□).

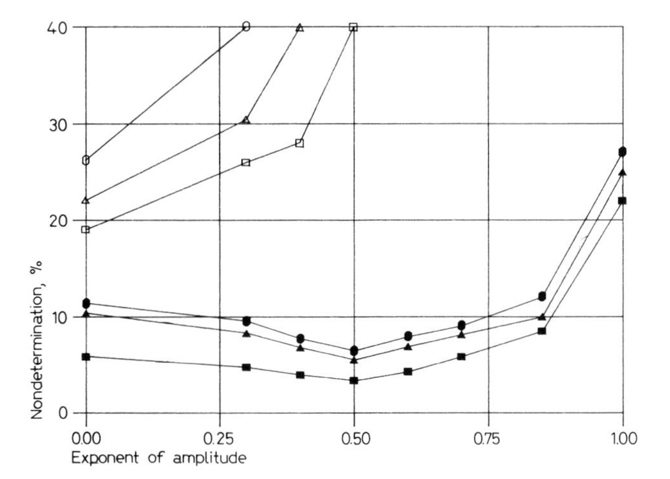

Fig. 7. Mean (n = 360) indices of nondetermination between the torque and (open symbols) the FITEMG and (closed symbols) the FDTEMG as a function of the coefficient a for m. brachialis (○), m. biceps brachii (Δ) and m. brachioradialis (□). The quantification was based upon 80-ms samples and the optimal filter (time constant 160 ms) was used.

results in a 52% improvement of the FIEMG over the IEMG. A side effect of the filter is that the optimal time delay is shifted to zero.

Finally the effect of the transformation of the EMG is illustrated in figure 7 for both the FDTEMG and the FITEMG. It seems that the relation between the static torque and the integrated EMG is disturbed when the influence of the amplitude is reduced. On the contrary, using a quantification of the EMG signal based on the differential technique, lower coefficients of nondetermination were found when the influence of amplitude was lowered. Using the FDTEMG with the obtained optimal value for a (a = 0.5), a significant mean improvement of 36% is found over the FDEMG.

Results of a Quadratic Regression

Both for the IEMG and for the DEMG, the standard errors of estimate obtained with a linear regression showed no significant differences with the standard errors of estimate obtained with a quadratic regression (t = 0.91 and 1.91, respectively). Furthermore, the coefficient of the quadratic term of the regression equation is small and may be both positive and negative.

Summary

In summary, then, a method of quantification has been developed that is based on three principles: (a) the reduction of the effects of intermittent signal amplitude changes by a normalisation procedure; (b) the use of differentiation rather than integration, and (c) the utilisation of low-pass filtering to capitalise on the effects of overlapping responses. The results of the application of these principles has been shown to be superior to the methods of integrating the EMG previously employed. In figure 8 there is persuasive evidence that the application of these principles increases the information extracted from the EMG. The progressive effects of the four recommended steps are illustrated and compared with the results of IEMG analysis.

The much more faithful predictions of muscle force that are associated with the FDTEMG, especially when short sample times are requested, may be seen in figure 9. One may conclude that the developed method results in highly accurate estimations of muscular activity which meet the specific needs of the analysis of human movement.

Application of the FDTEMG in the Analysis of Elbow Flexion

Determination of the Relative Contribution of Three Flexor Muscles

The exerted torque at the elbow is mainly the result of the force produced by three muscles: m. biceps brachii, m. brachialis and m. brachioradialis [Amis et al., 1979]. In order to determine the relative contribution of each muscle, the FDTGEMGs for each muscle were weighted with different sets of weights and then summed. These weights, which were always constrained to add up to ten, were sequentially altered from 0 to 10 in integer steps. Thus, 66 possible weighted combinations were available for examination.

An interpretation of these weight factors with regard to the relative contribution of the muscles is hindered because the values of the FDTEMGs do not only depend on the intensity of the muscular contraction, but also on the amplifier characteristics, the place, size and type of the electrodes and the skin resistance. In order to estimate the real activity of the muscle, a reference value is defined for each muscle in each test session. This reference value is the maximal FDTEMG registered under the same conditions and defined as 100% activity. The actual muscular activity is then expressed as a percentage of this maximal value.

With a sample time of 80 ms, a minimal nondetermination of 2.89% for the test was found with a weight factor of 40% for m. biceps brachii, 20%

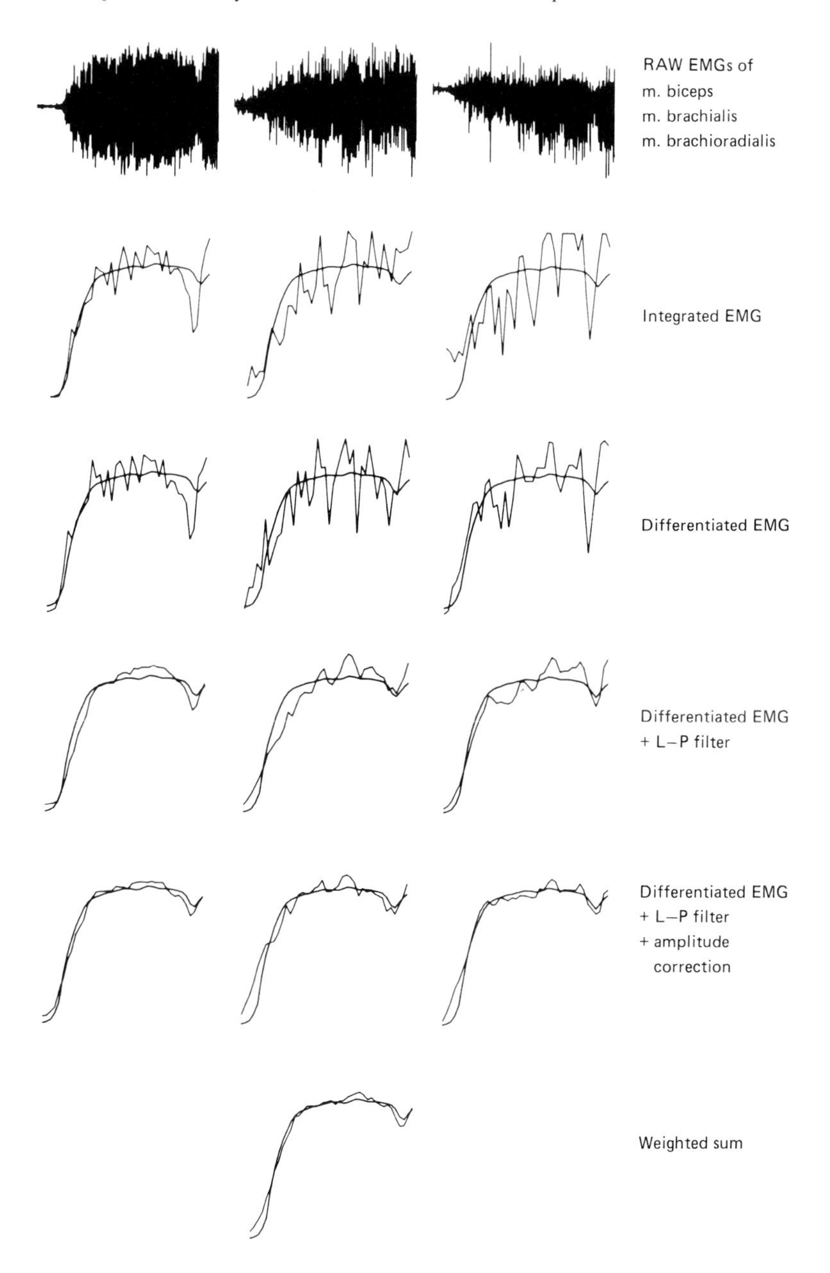

Fig. 8. Example of the different steps in the quantification of the raw EMGs of three elbow flexor muscles, using EMG samples of 80 ms during 3.6 s. The smoothed line is the static elbow flexion torque.

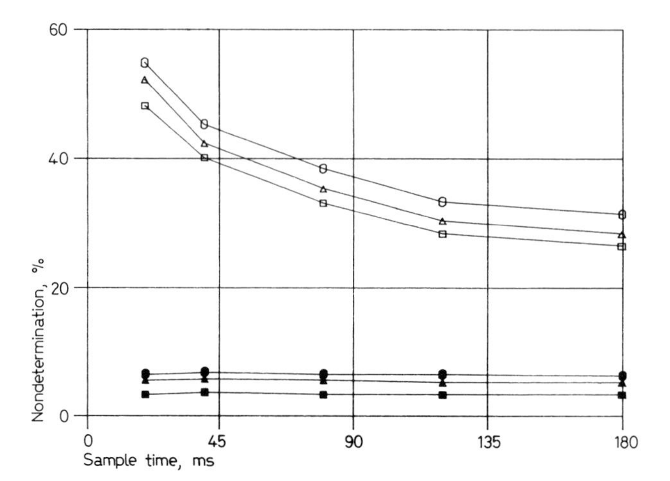

Fig. 9. Mean (n = 360) indices of nondetermination between the torque and (open symbols) the IEMG and (closed symbols) the FDTEMG as a function of the sample time for m. brachialis (o), m. biceps brachii (Δ) and m. brachioradialis (□). For the FDTEMG the optimal time constant and the optimal value of coefficient a was used.

for m. brachialis and 40% for m. brachioradialis. Relating only one muscle to the total moment exerted, significantly higher nondeterminations were found (fig. 9): 3,5% for m. brachioradialis, 5.6% for m. biceps and 6.5% for m. brachialis. For the retest a nondetermination of 2.81% was found with the same weight factors. These factors were independent of the angle at the elbow.

Interpretation of the Organisation of Elbow Flexion

Using the definition of muscular activity as mentioned above, figure 10 shows the degree of muscular activity of the three muscles at the instant of maximal torque during maximal isometric contractions at six different angles.

At an angle of 120 degrees the three muscles deliver an activity between 70 and 73% of their maximal values. At 175 degrees the activity of m. brachialis increases to 91.5%, while m. biceps brachii and m. brachioradialis show a decreasing activity to 61 and 59%, respectively. A reverse trend occurs at 75 degrees.

These findings may be related to the resulting force at the elbow joint. At an angle of 175 degrees, a contraction of both m. biceps brachii and m. brachioradialis results in relatively high axial forces at the elbow joint

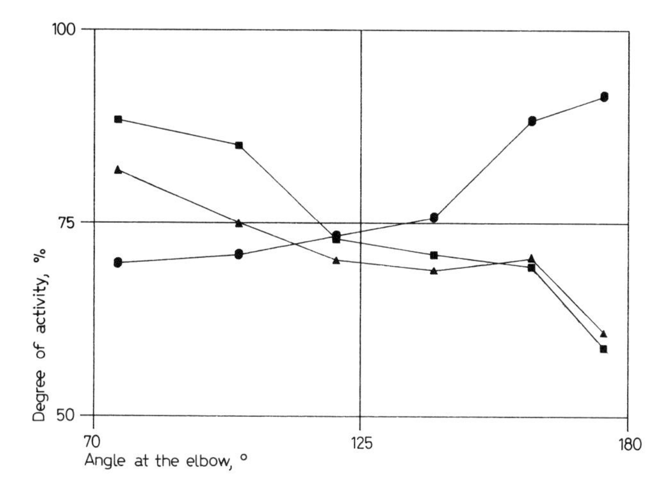

Fig. 10. Muscular activity (as a percent of the maximal value) of three muscles during maximal isometric contractions at different angles: (●) = m. brachialis; (▲) = m. biceps brachii; (■) = m. brachioradialis.

with regard to the limited flexion torque delivered. The m. brachialis, whose tendon inserts at a more favourable angle, delivers a relatively high flexion torque with a relatively low axial force component. The fact that at this angle the latter muscle is more active may correctly be interpreted as a limitation of the resulting force at the elbow joint. These findings suggest that it is not valid to assume that there is constant maximum muscular activity during maximum efforts.

Summary

The detailed analysis of human movement may depend upon accurate information being provided about the nature and extent of muscular activity. Electromyography is often used to provide this information. Unfortunately, as a review of the literature clearly reveals, the relationship between patterns or magnitudes of muscular activity and the electromyogram is neither as accurate nor as clear as is desirable. To meet this difficulty a new method has been developed called filtered, differenced and transformed electromyography (FDTEMG) which achieves significant improvements in the quality of EMG quantification. It is concluded that the method allows an accurate estimation of static elbow flexion torque, with an index of nondetermination of 2.9%, even when small EMG samples are used.

References

Amis, A.; Dowson, D.; Wright, V.: Muscle strengths and musculo-skeletal geometry of the upper limb. Engng. Med. *8:* 41–48 (1979).

Barnes, W.S.: The relationship of motor-unit activation to isokinetic muscular contraction at different contractile velocities. Phys. Ther. *60:* 1152–1158 (1980).

Basmajian, J.V.; Clifford, H.C.; MacLeod, W.D.; Nunnally, H.N.: Computers in electromyography, p. 138 (Butterworths, London 1975).

Benoit, J.; Hainaut, K.: Méthode d'étude quantitative de l'EMG global. Kinanthropologie *5:* 23–26 (1973).

Bigland-Ritchie, B.: EMG/force relations and fatigue of human voluntary contractions. Ex. Sport Sci. Rev. *9:* 75–117 (1981).

Bouisset, S.: EMG and muscle force in normal motor activities; in Desmedt, New concepts of the motor unit, neuromuscular disorder, electromyographic kinesiology, New developments in electromyography and clinical neurophysiology, vol. 1, pp. 547–583 (Karger, Basel 1973).

Corser, T.: Temporal discrepancies in the electromyographic study of rapid movement. Ergonomics *17:* 389–400 (1974).

Grieve, D.W.: Electromyography; in Grieve, Miller, Mitchelson, Paul, Smith, Techniques for the analysis of human movement, pp. 109–149 (Lepus Books, London 1975).

Hagberg, M.: The amplitude distribution of surface EMG in static and intermittent static muscular performance. Eur. J. appl. Physiol. *40:* 265–272 (1979).

Hayward, M.: Automatic analysis of the electromyogram in healthy subjects of different ages. J. neurol. Sci. *33:* 397–413 (1977).

Heckathorne, C.W.; Childress, D.S.: Relationships of the surface electromyogram to the force, length, velocity and contraction rate of the cineplastic human biceps. Am. J. phys. Med. *60:* 1–19 (1981).

Henneman, E.; Olson, C.B.: Relations between structure and function in the design of skeletal muscle. J. Neurophysiol. *28:* 581–598 (1965).

Hinson, M.; Rosentswieg, J.: Comparative electromyographic values of isometric, isotonic and isokinetic contraction. Res. Q. *44:* 71–78 (1973).

Hof, A.L.; Van Den Berg, J.W.: Linearity between the weighted sum of the EMGs of the human triceps surae and the total torque. J. Biomech. *10:* 529–539 (1977).

Inman, V.T.; Ralston, H.J.; Saunders, J.B. de C.M.; Feinstein, B.; Wright, E.E., Jr.: Relation of human electromyogram to muscular tension. Electroenceph. clin. Neurophysiol. *4:* 187–194 (1952).

Khalil, T.M.; Martinez, A.G.; Boykin, W.H.: Electromyographic models of mechanical torque; in Komi, Proceedings of the International Congress of Biomechanics 5, Jyvaskyla 1975. Biomechanics V-A; International series on biomechanics, No. 1, pp. 233–239 (University Park Press, Baltimore 1976).

Knowlton, G.E.; Hines, T.F.; Keever, K.W.; Bennett, R.L.: Relation between electromyographic voltage and load. J. appl. Physiol. *9:* 473–476 (1956).

Komi, P.V.; Buskirk, E.R.: Effect of eccentric and concentric muscle conditioning on tension and electrical activity of human muscle. Ergonomics *15:* 417–434 (1972).

Komi, P.V.: Relationship between muscle tension, EMG and velocity of contraction under concentric and eccentric work; in Desmedt, New concepts of the motor unit, neuromuscular disorder, electromyographic kinesiology, New developments in electromyography and clinical neurophysiology, vol. 1, pp. 596–606 (Karger, Basel 1973).

Komi, P.V.; Rusko, H.: Quantitative evaluation of mechanical and electrical changes during fatigue loadings of eccentric and concentric work. Scand. J. rehab. Med. suppl. *3, pp.:* 121–126 (1974).

Kuroda, E.; Klissouras, V.; Milsum, J.: Electrical and metabolic activities and fatigue in human isometric contractions. J. appl. Physiol. *29:* 358–367 (1970).

Lippold, O.C.J.: The relation between integrated action potentials in a human muscle and its isometric tension. J. Physiol. *117:* 492–499 (1952).

Magora, A.; Gonen, B.: Computer analysis of the relation between duration and degree of superposition of electromyographic spikes. Electromyog. clin. Neurophysiol. *17:* 83–98 (1977).

Maton, B.; Bouisset, S.: The distribution of activity among the muscles of a single group during isometric contraction. Eur. J. appl. Physiol. *37:* 101–109 (1977).

Norman, R.W.; Nelson, R.C.; Cavanagh, P.R.: Minimum sampling time required to extract stable information from digitized EMGs; in Asmussen, Jorgensen, Proceedings of the International Congress on Biomechanics 6, Copenhagen 1977. Biomechanics VI-A; International series on biomechanics, No. 6, pp. 237–243 (University Park Press, Baltimore 1978).

Norman, R.W.; Komi, P.V.: Electromechanical delay in skeletal muscle under normal movement conditions. Acta physiol. scand. *106:* 241–248 (1979).

Petrofsky, J.S.; Lind, A.R.: Frequency analysis of the surface electromyogram during sustained isometric contractions. Eur. J. appl. Physiol. *43:* 173–182 (1980).

Rau, G.; Vredenbregt, J.: EMG-force relationship during voluntary static contractions (m. biceps); in Cerquiglini, Venerando, Wartenweiler, International Seminar on Biomechanics 3, Rome 1971. Biomechanics III; Medicine and sport, vol. 8, pp. 207–274 (Karger, Basel 1973).

Rau, G.: Improved EMG quantification through suppression of skin impedance influences; in Nelson, Morehouse, Proceedings of the International Seminar on Biomechanics 4, University Park 1973. Biomechanics IV; International series on sport sciences, No. 1, pp. 322–327 (University Park Press, Baltimore 1974).

Rosentswieg, J.; Hinson, M.M.: Comparison of isometric, isotonic and isokinetic exercises by electromyography. Arch Phys Med Rehabil. *68:* 249–252 (1972).

Rothstein, J.; Delitto, A.; Sinacore, D.; Rose, S.: Electromyographic, peak torque and power relationships during isokinetic movement. Phys. Ther. *63:* 926–933 (1983).

Sato, B.: Functional characteristics of human skeletal muscle revealed by spectral analysis of the surface electromyogram. Electromyogr. Clin. Neurophysiol. *22:* 459–516 (1982).

Seliger, V.; Dolejs, L.; Karas, V.: A dynamometric comparison of maximum eccentric, concentric and isometric contractions using EMG and energy expenditure measurements. Europ. J. appl. Physiol. *45:* 235–244 (1980).

Sherif, M.H.; Gregor, R.J.; Liu, L.M.; Roy, R.R.; Hager, C.L.: Correlation of myoelectric activity and muscle force during selected cat treadmill locomotion. J. Biomech. *16:* 691–701 (1983).

Simons, D., Zuniga, E.: Effect of wrist rotation on the XY-plot of averaged biceps EMG and isometric tension. Am. J. phys. Med. *49:* 253–256 (1970).

Thys, H.; Vanderstappen, A.; Lhermout, C.: Stabilization of the isoelectric line of surface EMG by means of high pass filters. Electromyog. clin. Neurophysiol. *17:* 393–400 (1977).

Vigreux, B.; Cnockaert, J.C.; Pertuzon, E.: Factors influencing quantified surface EMG. Eur. J. appl. Physiol. *41:* 119–129 (1979).

Vredenbregt, J.; Rau, G.: Surface electromyography in relation to force, muscle length and endurance; in Desmedt, New concepts of the motor unit, neuromuscular disorder, electromyographic kinesiology. New developments in electromyography and clinical neurophysiology, vol. 1, pp. 607–622 (Karger, Basel 1973).
Winter, D.A.; Rau, G.; Kadefors, R.: Units, terms and standards in the reporting of myographical research. Int. Soc. Biomech. Newslett. *June*: 3–5 (1979).

Dr. M. Van Leemputte, Instituut voor Lichamelijke Opleiding, Katholieke Universiteit Leuven, Tervuursevest 101, B-3030 Leuven, Heverlee (Belgium)

Subject Index